The Research Impact Handbook

2nd Edition

Mark S. Reed

Fast Track Impact

Published by Fast Track Impact www.fasttrackimpact.com
St Johns Well, Kinnoir, Huntly, Aberdeenshire AB54 7XT

British Library Cataloguing-in-Publication Data
A catalogue record for this book is available from the British Library

This book should be cited as:
Reed, M.S. (2018). The Research Impact Handbook, 2nd Edition, Fast Track
Impact.

ISBN: 978-0-9935482-4-6 (pbk)
ISBN: 978-0-9935482-5-3 (ebk)

Design by: Anna Sutherland

To Hazel, Alfie and Isobel, in the hope that your incurable curiosity is never cured

In memory of Joanneke Kruijsen whose ideas inspired the paper this book is based on

Contents

Preface to the 2nd edition

Since the first edition of this book was published in 2016, it has been reprinted five times and used to train almost 5000 researchers from >200 institutions in 55 countries. As I have taken the book on the road, I have seen people's perceptions of impact transformed as they gain the confidence and skills they need to effect genuine lasting change through their work.

However, as I have conducted further research into the effects of the impact agenda, I have also noticed growing disquiet about some of the negative unintended consequences of incentivising researchers to generate benefits for society from their work. One civil servant told me "I'm fed up of being called up by researchers who want to have an impact on me when I've got a job to do". I have witnessed researchers telling stakeholders how they plan to "use" them to advance their academic careers.

Researchers interviewed for some of the papers I've published have explained how they have adapted the research they do as a result of the impact agenda, prioritising more applied work and in some cases compromising research quality. In countries that reward institutions for their impact, intrinsic motivations for working with publics and stakeholders are increasingly being 'crowded out' by extrinsic motives, leading in some cases to game playing. At its best, game playing gives undue credit for impact. At its worst, game playing uses publics and stakeholders as pawns in games of personal and institutional competition for scarce funding and reputational rewards, undermining public trust in the academy. The political roots of the impact agenda continue to fuel suspicion that it is an extension of neoliberal political agendas to marketise the academy, only valuing research in narrow, instrumental terms as a return on public investment. Combined with the rise in managerialism in the academy, some feel imperatives to achieve impact are yet another threat to academic freedom.

I believe that all of these concerns are valid, and they fuel my desire to get this book into even more people's hands. With this book, I want to appeal to people's intrinsic motives, and present a vision for an impact agenda with the needs of people and society at its heart. The roots of the impact agenda are far deeper and more ancient than any neoliberal plot to hijack impact for political ends. We need a new breed of grass-roots leaders, who are passionate about making a difference for the right reasons, and who can inspire others to follow in their footsteps. Impact, ultimately, is about long-term, trusting, two-way relationships. Some of these relationships might lead to impacts, but others might not. The point is that researchers need to be in this for the long haul, rather than dropping partners as soon as the project is over, or as soon as it becomes apparent that it won't help generate impacts we can report.

In this second edition of *The Research Impact Handbook*, I want to reconnect researchers with the concept at the heart of the impact agenda: empathy. If we don't do this, wider society's cynicism of research and researchers will only grow further. If we do, we have the opportunity to achieve change on an unprecedented scale. Put simply, research impact is the good that researchers can do in the world. What is the good you can do?

I hope that this book continues to motivate and inspire a new generation of researchers who see impact as a privilege rather than a duty, and who seek impact because they want to, not because they are told to. As my friend and colleague Ana Attlee once advised me: *"Every night without fail, ask yourself, what did I do well today, how did I make a difference, where did I leave an impact?"*

What's new in the 2nd edition?

Much has changed since the first edition was published less than two years ago. First, the impact agenda has become significantly more prominent in the global research community, both in terms of directing research funding and assessing research excellence. The majority of national research funding organisations around the world now require researchers to make some sort of statement about the likely impact of the work they are seeking funding for. Australia now assesses research impact in its Engagement and Impact Assessment as part of Excellence in Research for Australia, and the European Commission is looking at how to evaluate impact more effectively in the successor to its Horizon 2020 funding programme. The forerunner to these schemes is the UK's Research Excellence Framework, which included an assessment of the significance and reach of impacts for the first time in 2014. In the 2021 iteration of this exercise, 25% of scores (and hence funding) given to Higher Education Institutions will come from impact (the rest coming from research outputs and research environment).

Second, there is now a significantly richer evidence base upon which to draw, new tools are available and new best practices are emerging as the research community learns how to engage more effectively with stakeholders and publics. In this second edition, I have tried to integrate as much as possible of the latest and most important evidence, tools and experience that have become available:

- I have added a new chapter defining impact in more detail, while keeping it as simple as possible, explaining how impact works and describing the different types of impact that are possible.

- In Chapter 4, I have added a section on what to do if you think your research might make money, briefly covering the basics of intellectual property.
- As the most important of the five principles underpinning my relational approach to impact, I have significantly expanded Chapter 6 which deals with the 'engage' principle.
- In Chapters 4–8, I have explained how each principle can help you achieve meaningful and lasting impacts in more time-efficient ways.
- Chapter 10 now includes a section on my best practice library of pathways to impact, and more detailed guidance on how to write the impact sections of a grant proposal (including common mistakes researchers make).
- Chapter 14 has been revised to include more on prioritising publics as well as stakeholders, and introduces interest-benefit matrices in addition to the more traditional interest-influence matrices covered in the first edition.
- As you might expect, much has changed in the digital world in the last couple of years, and so Chapter 17 has been extensively revised and updated to include more material on things you can do without having to engage with social media to drive impacts online. Chapter 18 on using Twitter and LinkedIn has also been updated.
- I've added a new chapter on presenting with impact because how we deliver our message can significantly influence whether or not we influence our listeners and achieve impact. I received voice coaching a few years ago and wish I'd known what I learned many years earlier. The material in Chapter 19 will transform how you speak, so you can transform your listeners.
- I've made a number of changes to Chapter 20, including adding a section explaining how my relational approach to evidence-informed policy better represents the reality of real-world policy processes than outmoded conceptions of evidence-based policy.
- In Chapter 21, I have explained how I 'stress-test' policy briefs with people from opposing sides of a policy

debate to identify weaknesses and fix them as part of my relational approach to writing policy briefs.

- Chapter 22 on tracking, evaluating and evidencing impact has been completely rewritten and significantly expanded based on an extensive review I've undertaken of the latest literature on this topic (forthcoming in the peer-reviewed literature).
- Lastly, in the final part of the book, I've created a new series of short 'how to' guides and worked examples covering:
  - Getting testimonials to corroborate the impact of your research
  - Writing up an impact evaluation as a research article
  - Evidencing international policy impacts
  - How to set up a stakeholder advisory panel for your research project
  - How to turn your next paper into an infographic
  - How to write a top-scoring impact case study for the UK Research Excellence Framework
  - How to crowdfund your next project
  - How to turn your research findings into a video that people actually want to watch
  - Publics/stakeholder analysis: worked example
  - Event facilitation: example plan

Chapter 1
Introduction

"No matter what people tell you, words and ideas can change the world."

Robin Williams

"An idea that is developed and put into action is more important than an idea that exists only as an idea."

Buddha

"Good ideas are not adopted automatically. They must be driven into practice with courageous patience."

Hyman Rickover

Imagine what might be possible if we could harness the collective wisdom of the world's most intelligent people to tackle the challenges facing the world today. We could do amazing things.

Researchers are under more pressure today than ever before to demonstrate the economic and social benefits, or 'impact', of their work. But we have been trained how to do research, not how to generate impact. This means many of us feel unprepared and out of our depth when we think about working with people who might be interested in our research. It is hard to know where to start. Putting out a press release doesn't usually do much. Even if the story is taken up widely, knowing how to convert media interviews into real economic and social benefits is a whole different thing.

If you have no idea how your research could make a difference, then this book will help you identify practical things you can do to start on a journey towards impact.

If you know the sorts of impact you would like your research to have, but don't have the confidence, skills or ideas to make it happen, this book will give you the tools to move forward with.

If you are already making a difference and want to take your impact to a new level, this book will help you refine your practice and become more efficient as a researcher, so you've got more time to have an even greater impact.

You may feel daunted by the challenge, but I want to share with you how straightforward (and fun) it can be to embed impact in your research. The principles and steps I will show you in this book will be just as effective whether you're researching medical microbiology or mediaeval monasteries; whether you are a PhD student or a professor. They are based on research I have done with colleagues over the last decade into the way we generate and share knowledge, and they have been tried and tested by researchers I have trained around the world.

For many of these researchers, simply understanding how they can achieve impact from their research is a revelation. They discover a new sense of motivation, knowing that the paper or book they are writing won't just end up on a library shelf gathering dust, but will actually be put into practice or change public perceptions. They feel empowered when they discover the new digital tools that can help them reach out to more people than ever before. For many of the researchers I work with, just discovering how many people are genuinely interested in their work creates a thirst to understand how they can engage with these people more effectively. I think

people are often surprised at how quickly they can start to realise impacts once they understand a few basic principles and start taking clearly-thought-out, deliberate steps towards impact.

What is stopping you putting new knowledge into practice?

We live in an increasingly networked world in which academic collaborations are growing in size and disciplinary diversity. However, there remain many important barriers that prevent us from putting our knowledge into practice. Putting aside issues of funding, politics and other external barriers for a moment, I believe that the greatest barrier is *us*.

When I train researchers, there are two things that come up time and again. The first is that researchers are trained how to generate and test new ideas, but not how to communicate them or put them into practice. Yet researchers are increasingly expected by governments and other research funders to demonstrate tangible impacts from their work. The second issue is that many researchers are genuinely intimidated by the prospect of doing research *with* rather than *to* or *for* people who are interested in their work. It is hard enough working out who might be interested in our work, let alone having the confidence to actually connect with them.

Of course, it is easier for some of us to generate impact than others. A researcher once told me, pointing at a picture I'd shown him of a sunset over a bog pool, "it's easy for you — you've got a sexy research topic". Well, that was the first time I'd heard that particular adjective used about a peat bog. Although I love bogs, I have to admit that they do have a bit of a public image problem. So, I was pleased that the social media campaign that the image came from was convincing someone. Even if you imagine that people will perceive your research as unintelligible or boring, there is usually someone somewhere who cares about some part of what you do, or there is some way of making it interesting or useful. Sometimes it is a journey you can take in a few simple steps, and sometimes it is one that takes years. However, to date I have yet to meet a researcher who doesn't find the journey incredibly rewarding, no matter how long it takes.

For some researchers, it is the final goal that motivates them — changing policy or practice, licencing their patent, or changing public perceptions. For many, the journey itself ends up being just as rewarding. I know lots of researchers who grudgingly started engaging with people who were interested in their research, and who then discovered that those relationships led to new collaborations and funding. Those projects led to new discoveries, which in turn fuelled the sense of curiosity that first brought them into research. One project I led ended up bringing in as much funding from industry as the original grant, and that enabled us to increase our sample size and do more rigorous research. It eventually led to a whole new body of research that we couldn't have envisaged when we started the work, which has become career-defining for me. Of course, there are researchers who don't believe that we should have to justify our existence by evidencing our impact. And of course, pure, non-applied research has equal value and is vitally important. But whether we like it or not, those who fund research (mainly taxpayers) increasingly want to see tangible impacts from it.

I think it is easy to forget the privileged position we are in as researchers. We have the latest research insights and evidence at our fingertips and (crucially) we have the knowledge to be able to critically interpret and use what we learn. Yet, without realising it, we hide these insights in impenetrable language, on inaccessible bookshelves, out of reach of those who need our knowledge most. The American women's rights advocate Margaret Fuller (1810–1850) wrote, "if you have knowledge, let others light their candles from it".

With the advent of open access publishing, research is increasingly available online. However, available research isn't necessarily accessible research. We still have a tendency to use language to erect walls around our candles, so others outside academia or our own discipline can't see the flame, let alone light their own candle from it. We need to take down the walls, one piece of jargon at a time, if we want to communicate our research effectively.

But often that is not enough. We still have to draw people close enough to our work to actually see the insights, appreciate their relevance, and turn them into knowledge they can use. Typically, this means we have to cradle the flame and carry it to people, rather than just hope that people will be drawn to the light. Rather than

Part 1: Principles to underpin your impact

Chapter 2
What is impact?

Summary

This chapter defines research impact and explains how impacts are generated through knowledge exchange. Put simply, research impact is the good that researchers can do in the world. It consists of the non-academic benefits that arise, whether directly or indirectly, from research. Knowledge exchange is a precursor to impact, and this happens through learning, when the data and information from research become knowledge that people can benefit from or use. There are many factors that can influence the likelihood of research leading to impact, including the context you are working in, who is involved and how, your approach to knowledge exchange and how well you manage power dynamics.

What is impact?

There are many different ways of thinking about impact, and this chapter will describe some of the most commonly used approaches and definitions. The word 'impact' is problematic for many, given its connotations of (possibly painful) collisions. For me, however, it is quite simple. Impact is the good that researchers can do in the world.

There is an implicit value judgment in this definition; we are seeking benefits and working for the good of others beyond the academy. This means we need to reflect on whether there may also be unintended negative consequences, and do everything we can to avoid these. It is our responsibility as researchers to anticipate and assess the potential consequences of research and work with stakeholders to design responsible, sustainable and inclusive research (for more information, see resources on 'responsible research and innovation' under 'further reading' at the end of this book). With this in mind, we can simply define impact as benefit. It is surprising how much clarity it brings, when you simply ask yourself "What was the benefit?". Keep asking who benefits and how, and you will find impact. People regularly ask me when does the 'pathway to impact' stop and the impact begin? The answer is that impact starts when you see benefits.

There is an implicit venue for those benefits in my definition: they lie beyond the academy. There are, of course, many forms of academic impact we may be equally interested in (for example, bibliometric indicators of impact), but I am concerned in this book with non-academic impacts.

Impact may be direct or indirect. If someone else is able to use your non-applied research (say a new mathematical algorithm or theory) to derive significant benefits (say a piece of software that saves lives), and that benefit would not have been possible without your research, then you can share some of the credit for that impact.

Of course, for this to be 'research impact', the benefits must be clearly linked to your research. This doesn't mean that every part of your work needs to be used. Things can go wrong when people cherry pick the parts of your work that they like and overlook parts that are uncomfortable for them. However, very often only one of your findings is relevant for a particular group, or someone might be interested in the theory or method behind your work rather than the ultimate findings. It is also perfectly normal to go beyond your own research to draw on other evidence to help the people you are working with, or just get involved in some other way that has nothing to do with research but that helps make a difference. If you are drawing on other people's research, that's still research impact (but you won't be able to claim this as impact from your research). If you are doing something else to help that is not related to research, then that's still impact, but it isn't research impact (and you won't be able to claim that as impact from your research either). It is important to be prepared to 'go the extra mile' and help those you are working with in ways that go beyond your own research if you want to maintain trust and avoid the perception that you are only doing this for your own gain. In many cases, the most effective approach is to find other researchers who can help. In this way, you are able to add value to the publics and stakeholders you are working with, whilst providing opportunities for impact to your colleagues.

Finally, impact is often conceptualised as beneficial change, but we may have just as much of an impact if our research prevents a damaging or harmful change from occurring. Impacts can be immediate or long-term, in our back yard or in outer space, transforming one person's life or benefiting millions, tangible or elusive. Defined broadly, impact is rich and varied, and has value whether or not you are able to 'prove' it to others. However, if you want to robustly claim and talk publicly about the impact of your research, the impacts will need to be demonstrable. There are two ways in which you will need to demonstrate impact: you will need to provide evidence that you achieved impact (and ideally that this was significant and far-reaching); and you will need to provide evidence that your research contributed toward achieving those impacts. The key word here is 'contribution'. It is rare that a researcher is able to claim all the credit for an impact linked to their work. There are almost always other lines of evidence (or argument) that have contributed toward the eventual impact. I'll discuss issues around attribution in Chapter 22. The need to demonstrate impact tangibly may skew researchers towards particular types of impact that are easier to attribute to the research and evidence.

That's why as a result, definitions of research impact from institutions charged with assessing impact may include demonstrability. It is not enough just to focus on activities and outputs that promote research impact, such as staging a conference or publishing a report. There must be evidence of research impact, for example, that the report has been used by policy-makers, and practitioners, or has led to improvements in services or business. That's why Research Councils UK defines research impact as "the *demonstrable contribution* that excellent research makes to society and the economy" (italics added).

How good is my impact?

I will look at evaluating and evidencing impact further in Chapter 22, but it is worth noting here that impact is usually judged against two criteria: significance and reach. First, ask yourself how significant the benefits of your work are. How meaningful, valuable or beneficial is your work to those you are working with? Second, ask yourself how far-reaching your work is. Are there other groups who might benefit in similar ways, or new applications of your work that could bring new benefits to new groups?

For me, the order in which you ask yourself these two questions is crucial. I would argue that if you do something that is situated in every country of the world across multiple social groups, but no one really cares, or benefits in any tangible or meaningful way, you don't actually have an impact. On the other hand, if you save one person's life as a result of your research, you clearly have a significant impact. Therefore, first ask yourself what you can do that would be significant on whatever scale you feel is achievable to you at this point. It may be one company, your local community or your local hospital, but if you think you could actually achieve something significant on that scale, then focus on that.

Don't be put off by others who have gone before you and achieved impacts on grand scales. Start small, and once you can see that it works (and only then), ask how you might be able to extend the reach of your impact. At this point, ways of extending your reach are often self-evident, and you will have met people who will want to help you on this journey. Moreover, you have the evidence that it worked for one company, one community or one hospital, which makes it much easier for others to follow in their footsteps.

What types of impact are there?

There are many different types of impact, with some types leading to others. Institutional definitions of impact often list types of impact, but there have been few attempts to categorise these to date. For example, the Higher Education Funding Council for England defines impact as "an effect on, change or benefit to the economy, society, culture, public policy or services, health, the environment or quality of life, beyond academia". More simply, the Australian Engagement and Impact Assessment defines impact as "the contribution that research makes to the economy, society, environment or culture, beyond the contribution to academic research".

Based on my own analysis of impact case studies from around the world, I distinguish between ten types of impact (see Chapter 22 for ideas about how you might evaluate each of these different types of impact). Categorising impacts in this way is useful, because it gives you a checklist for considering the full range of possible impacts you could seek. Even if you have a narrow focus on one type of impact (say economic impact of a spin-out company), it is often worth looking through the other types of impact that might arise, to consider whether you might also be generating these benefits. For example, a company's new product may replace something that was energy intensive to produce, and so reduces greenhouse gas emissions, giving you an environmental impact as well as the original economic impact.

Table 1 shows you the types of impact you can look for. The following subsections define each of these types of impact and give you examples of the sorts of things you might seek to do to achieve each type of impact.

Table 1: Research impact typology

Type of impact	Definition
Understanding and awareness	People understand an issue better than they did before, based on your research
Attitudinal	A change in attitudes, typically of a group of people who share similar views, towards a new attitude that brings them or others benefits
Economic	Monetary benefits arising from research, either in terms of money saved, costs avoided or increases in turnover, profit, funding or benefits to groups of people or the environment measured in monetary terms
Environmental	Benefits from research to genetic diversity, species or habitat conservation, and ecosystems, including the benefits that humans derive from a healthy environment
Health and well-being	Research that leads to better outcomes for the health of individuals, social groups or public health, including saving lives and improving people's quality of life, and wider benefits for the well-being of individuals or social groups, including both physical and social aspects such as emotional, psychological and economic well-being, and measures of life satisfaction
Policy	The contribution that research makes to new or amended laws, regulations or other policy mechanisms that enable them to meet a defined need or objective that delivers public benefit. Crucial to this definition is the fact that you are assessing the extent to which your research made a contribution, recognising that it is likely to be one of many factors influencing policy. It also goes beyond simply influencing policy, to enabling those policies to deliver

	public benefits. If the policy intervention would have had the same impact without the elements based on your research, can you really claim to have had impact? Arguing for the significance of your contribution is therefore an essential part of demonstrating that your research achieved policy impacts.
Other forms of decision-making and behaviour change impacts	Whether directly or indirectly (via changes in understanding/awareness and attitudes), research can inform a wide range of individual, group and organisational behaviours and decisions leading to impacts that go beyond the economy, environment, health and well-being or policy.
Cultural	Changes in the prevailing values, attitudes, beliefs, discourse and patterns of behaviour, whether explicit (e.g. codified in rules or law) or implicit (e.g. rules of thumb or accepted practices) in organisations, social groups or society that deliver benefits to the members of those groups or those they interact with
Other social	Benefits to specific social groups or society not covered by other types of impact, including, for example, access to education or improvements in human rights
Capacity or preparedness	Research that leads to new or enhanced capacity (physical, financial, natural, human resources or social capital and connectivity) that is likely to lead to future benefits, or that makes individuals, groups or organisations more prepared and better able to cope with changes that might otherwise impact negatively on them

1. Understanding and awareness impacts

Definition: people understand an issue better than they did before, based on your research.

For example, if you are doing public engagement, this may be a new appreciation for something people had previously overlooked or taken for granted, or you may have raised awareness about an important issue that typically gets limited media coverage. If you are working with stakeholders you may, for example, have done research that uncovers the scale and urgency of a problem which needs to be solved, or you may have evidence that a product or policy would have negative unintended consequences if introduced. You may not have the solution to these problems, but the fact that people are now aware of the issue is an impact in itself. It is worth noting here that awareness and understanding impacts often lead (in time) to other types of impact (below), so you may want to wait until these other impacts have occurred before reporting your impacts to funders and assessors, who are likely to be more interested in what eventually happened as a result of this new understanding.

2. Attitudinal impacts

Definition: a change in attitudes, typically of a group of people who share similar views, towards a new attitude that brings them or others benefits.

For example, public engagement might lead to a new appreciation for alternative views and more positive perceptions of people who hold differing views. This would be a benefit if it led to less prejudice towards others. An example of this might be a research intervention that led to measurable changes in racist attitudes in a sporting community, such as football. Alternatively, you might change attitudes of white male business executives towards female or non-white employees seeking board level positions in corporations. Again, changes in attitudes often lead (in time) to changes in behaviour and decisions (e.g. leading to more non-white footballers or company board members), and you may want to design a pathway to impact that seeks attitudinal change on the way to other impacts (below) before reporting the non-academic benefits of your research.

3. Economic impacts

Definition: monetary benefits arising from research, either in terms of money saved, costs avoided or increases in turnover, profit, funding or benefits to groups of people or the environment measured in monetary terms.

For example, public engagement based on your research may have led to a reduction in the number of people visiting their local doctor with a particular complaint, saving the health service, insurance providers or patients money. Alternatively, your work might have led to a new product or service that has made money for a company (or its shareholders) or saved a company money (increasing its profits). Your research might have demonstrated that a new or existing policy is not achieving its goal and so is wasting money, leading to the withdrawal of that policy (saving money) or its replacement with something that works (providing better value for taxpayers' money). You may have quantified the economic benefit to society of a new policy based on your work, or estimated the benefits of an intervention based on your work for the natural environment using monetary valuation techniques. Economics also increasingly offers a range of non-monetary valuation methods for assessing impacts from research that cannot easily (or should not) be converted to monetary values (more on that in Chapter 22).

4. Environmental impacts

Definition: benefits from research to genetic diversity, species or habitat conservation, and ecosystems, including the benefits that humans derive from a healthy environment.

Environmental benefits may be for nature alone (with no tangible benefit for people), or for nature and people. Research that only benefits nature (e.g. saving a species from extinction) is just as valuable as research that also then benefits people as a result of the benefits for nature (e.g. via health benefits from reduced pollution or increased well-being from access to green space). Alternatively, research may lead to human behaviour changes that benefit nature (e.g. reducing consumption or using less plastic). In this case, you ideally want to know whether people's behaviours changed, and also whether those changes actually had the environmental benefits you hoped for. You might try and evaluate environmental benefits for people in terms of understanding and awareness benefits (above) through education based on your work, or via the services that nature provides to people. Many of these services are tangible and easy to measure, such as the provision of food and health benefits, but many are less tangible and present challenges for measurement, such as cultural or spiritual benefits arising from interaction with nature. There is a growing range of approaches from social sciences and the arts and humanities to assess these sorts of impacts, which are covered in Chapter 22.

5. Health and well-being impacts

Definition: research that leads to better outcomes for the health of individuals, social groups or public health, including saving lives and improving people's quality of life, and wider benefits for the well-being of individuals or social groups, including both physical and social aspects such as emotional, psychological, economic well-being and measures of life satisfaction.

For example, research from disciplines such as clinical medicine, allied health, public health and biomedical sciences may reduce mortality and morbidity in certain patient groups, via interventions such as new drugs and treatments for diseases and conditions or public health interventions to shift individual behaviours towards more healthy outcomes. The applied nature of such research requires evidence of the effectiveness of interventions (their impact) as part of the research process, for example, through clinical trials and meta-analyses of multiple studies in different contexts. There is a wide range of research that may enhance well-being in other ways, including, for example, engaging women affected by domestic violence in reading groups based on research into 19th-century feminist literature as a way of supporting and empowering this group. Alternatively, research on the impact of students on communities around universities has led to many institutions investing in purpose-built student accommodation in different areas, enhancing the well-being of their local communities through reductions in things like late-night noise and litter.

Definition: the contribution that research makes to new or amended laws, regulations or other policy mechanisms that enable them to meet a defined need or objective that delivers public benefit. Crucial to this definition is the fact that you are assessing the extent to which your research made a contribution, recognising that it is likely to be one of many factors influencing policy. It also goes beyond simply influencing policy to enabling those policies to deliver public benefits. If the policy intervention would have had the same impact without the elements based on your research, can you really claim to have had impact? Arguing for the significance of your contribution is therefore an essential part of demonstrating that your research achieved policy impacts.

For example, your research may have been one of hundreds of studies in a particular area, but your work provided a missing link or some other crucial piece of evidence that made a policy possible. Alternatively, you may simply have been the person who was able to advise those developing the policy. You may have signposted your work alongside other key pieces of evidence and your evidence-based advice became crucial to the development of that policy. This would be research impact as long as your advice was based on research (ideally including your own), and this advice informed and shaped policy in ways that enhanced the policy, enabling it to deliver benefits more effectively. As a result, your research may have been heard in committees and cited in policy documents, but if it wasn't, you should be able to collect testimonials from members of the policy community explaining your role in the process, and the significance of your research. Attributing policy impacts to individual research projects or researchers is one challenge, but others include the significant time lags that exist between the production of evidence and policy influence, and ideological barriers to the uptake of certain types of evidence at certain times. These issues will be explored in greater detail in Chapter 20, which describes my relational approach to working with policy-makers.

Definition: whether directly or indirectly (via changes in understanding/awareness and attitudes), research can inform a wide range of individual, group and organisational behaviours and decisions, leading to impacts that go beyond the economy, environment, health and well-being or policy.

For example, your research may change the attitude of a particular group towards others (e.g. attitudes towards non-white footballers or board members of companies), and that attitudinal change may then translate into changes in behaviour in sports training and selection leading to more non-white footballers, or the promotion of more non-white employees to board positions. The impact of these behaviour changes may then lead to other impacts over time, such as improved performance of the football team or company, which you might then claim as an additional economic impact arising from the research. In addition to policy decisions (above), there are many other organisations who seek to base decisions on evidence from research, ranging from charities and non-governmental organisations to farmer co-operatives and arts organisations. Where you can demonstrate that your research has contributed significantly towards decisions that have delivered benefits for these organisations, you have achieved impact.

8. Cultural impacts

Definition: changes in the prevailing values, attitudes, beliefs, discourse and patterns of behaviour, whether explicit (e.g. codified in rules or law) or implicit (e.g. rules of thumb or accepted practices) in organisations, social groups or society that deliver benefits to the members of those groups or those they interact with.

For example, research on a classical composer may provide cultural impacts by opening up that composer's work to new audiences through interpretation (via public lectures, media work or pre-concert talks) and influence how performers interpret, perform and record the composer's work, leading to critical acclaim and enriching the cultural experience of the music-loving public. Alternatively, research on working-class entertainment might lead to changes in attitudes towards historic entertainment venues that had been left to fall into disrepair, leading to them being valued more greatly by members of the public. This might then lead to other forms of impact, for example, economic impacts based on restoring historic entertainment venues that bring in visitors and revenue to previously overlooked locations. Evidencing the cultural impacts can be challenging, but methods do exist (more in Chapter 22).

9. Other social impacts

Definition: benefits to specific social groups or society not covered by other types of impact.

For example, your research on micro-grids for solar energy might enable communities living in remote parts of Africa to access electricity, enabling school children to have access to artificial light so they can do homework. Your research on education in collaboration with researchers working on low-cost internet-connected devices might lead to the development of self-taught courses that give millions of children access to education that would otherwise not have been possible. Your research on the relationship between Islamic law and international human rights might lead to legislative reform that leads in turn to the empowerment of groups that had previously been discriminated against. Access to education and other human rights may not represent a cultural change, and may not lead to benefits for health and well-being, enhanced livelihoods or other economic measures or the environment. However, these other social impacts must also be recognised.

10. Capacity or preparedness

Definition: research that leads to new or enhanced capacity (physical, financial, natural, human resources or social capital and connectivity) that is likely to lead to future benefits, or that makes individuals, groups or organisations more prepared and better able to cope with changes that might otherwise impact negatively on them.

For example, your research might lead to the development of new infrastructure or equipment that will provide benefits to future users.

Your evidence may have enabled a charity to gain significant new funding for its future work. Your research may have led to the creation of new habitats that will protect critical infrastructure from future flooding. Key people may have new knowledge and skills that will enable them to generate impacts for their business or adapt to a change in the law. As a result of collaborating in your research project, people may be connected to a more diverse group of people or organisations who they now know and trust, and as a result they know who to contact and what capabilities they can deploy to respond faster and more effectively to a natural disaster. Many research funders specifically target these sorts of impacts, but others would view capacity or preparedness as simply a pathway to other impacts that are only realised once that capacity is used. As a result, you may want to wait and see how the new capacity and preparedness that results from your research is used. The problem is that this may take time, and in some cases the eventuality for which you helped people prepare does not come to pass.

In some countries, pedagogical impacts are also included as part of research impact. These can be important and far-reaching, but to avoid confusion between benefits for the academy and non-academic impacts, I have not included them in this typology.

How does impact happen?

You should now have a clearer understanding of what impact is, and the types of impacts that typically occur. In the rest of this book, I want to show you how you can achieve each of these different types of impact (and demonstrate that you achieved them). Depending on how interested you are in the process of impact generation, you may simply want to move onto the next chapter, and start thinking about how to generate impact. However, many of the researchers I work with are curious about the mechanisms and theory that explain how impacts happen. As there is limited literature on this subject, I am going to use the rest of this chapter to explain how impacts happen.

There is one universal precursor to impact: learning. The research we publish is typically in the form of data and information (useful data), but for someone to benefit or use your work, this data and information need to be transformed into knowledge in someone's head. This happens through learning: someone somewhere needs to learn about your work. Therefore, if you want your research to have an impact, you need to find new ways of making your work both accessible and understandable to the people who can benefit from or use your work most.

There are many overlapping terms that are often used to describe this process, including knowledge management, sharing, co-production, transfer, brokerage, transformation, mobilisation and translation. Each of these terms is used in different disciplinary or sectoral contexts to mean slightly different things. Some imply a one-way flow of knowledge from those who generate it to those who use it, whereas others imply different levels of two-way knowledge exchange and joint production of knowledge between those who need to use knowledge and researchers. For simplicity, in this book I will use the term 'knowledge exchange' to include all of these different approaches.

There are five factors that can explain whether or not a pathway to impact is likely to work:

1. **Context and purpose:** the impact generation process always starts in a given context, for example, the culture, educational status and interests of a particular public, or the emergence of a new challenge such as a new disease or opportunity such as a

new technology. Within this context, researchers and various social groups may wish to achieve specific benefits (your purpose or impact goal), for example, learning about the work of a nationally significant artist, or finding a cure for a disease. As contexts or purposes change over time, you need to adapt your pathway to impact, considering how you may deal differently with each of the factors below. For any given context and purpose, each of the steps required to generate impact will vary significantly.

2. **Who initiates and leads on the pathway to impact:** researchers, publics and/or stakeholders may initiate and lead the impact generation process. Who initiates and leads the process matters: there is evidence that impacts vary systematically based on the group that has ownership of the pathway to impact. For example, your pathway to impact may be self-organised from the bottom-up, initiated and led by those seeking the benefits. Alternatively, impact may be initiated through more top-down approaches, where plans to achieve benefits are initiated and led by researchers or other external agencies, such as the government.

3. **Representation:** your engagement with stakeholders and publics is likely to vary from full to partial representation of different groups and their interests. Partial representation may be deliberate (for example, as part of a phased approach to engaging increasingly influential or hard-to-reach groups), or due to a lack of time or resources. There is evidence that pathways to impact are significantly affected by who is engaged in the pathway, and inadvertently overlooking important groups can undermine your attempts to achieve impact.

4. **Design:** the way you engage with publics and stakeholders may be designed as communicative (one-way flows of knowledge from researchers to stakeholders and/or publics), consultative (one-way from stakeholders to researchers), deliberative (two-way knowledge flows) or co-productive (joint production of knowledge). Your choice of approach should be adapted to who you are engaging with (point 3 above), who initiated and is leading the process (point 2), and your context and impact goals (point 1).

5. **Power:** finally, depending on the design of the process and its facilitation, power dynamics between researchers, publics and stakeholders may be more or less effectively managed, strongly

influencing the ultimate achievement of benefits or unintended consequences.

Ultimately, the likelihood of your pathway to impact working depends on each of these five factors. Get these right, and you are highly likely to achieve your impact goals. Get them wrong, and you are far more likely to fail, potentially leading to unintended negative consequences.

The power of clear thinking about impact

Many of us think we already know what impact is, but in my experience even a short discussion about impact raises multiple questions and challenges. This often reveals incomplete or muddled thinking, which can be problematic when we need to think on our feet and adapt to changing circumstances in order to keep our impacts on track. By defining it in a way that is both incredibly simple and yet nuanced, I hope to have clarified your thinking on impact. It isn't complicated. It is simply the good that you can do in the world. By applying this definition to as many different contexts as possible, I hope that you now have a much clearer idea of the specific types of impact you might want to pursue. The examples I have given of each type of impact are far from exhaustive, but you should now have some concrete ideas of the sorts of things that could be considered research impacts. Finally, I have sought to explain the processes through which impact occurs. By explaining the various factors that influence whether or not you are likely to achieve impact, I hope to have empowered you to design your pathway to impact in a way that is highly likely to actually work. I want to move now from theory to practice, and draw on research I did a few years ago to distil 'what works' in impact from interviews with researchers, publics and stakeholders involved in pathways to impact. As a university professor myself, I know how little time researchers have, and so I have focused on explaining how each of the principles that emerged from this work can help you save time and fast track your impact.

Chapter 3
Five principles to fast track your impact

Summary

This chapter introduces five evidence-based principles that underpin successful knowledge exchange and impact. Based on the experiences and views of researchers who have worked with stakeholders and members of the public in projects around the world, the chapter provides many practical suggestions for ways you can increase your impact, whilst providing a conceptual framework for the rest of the book. If you want to have an impact, you need to embed knowledge exchange in your research. The five principles that can help you do this are: design impact into your research from the outset; represent all relevant interests in your work; engage with those who are interested in your work to build long-term, two-way, trusting relationships; achieve early impacts to build credibility and motivate engagement; and reflect on your knowledge exchange practice so you can sustain impacts for the long term.

I still remember vividly the day that the idea occurred to me. If I took the time to explain it, you probably wouldn't be that impressed, as it is so obvious. But it is the only time I can claim to have had anything close to a 'eureka' moment.

I remember standing on the stairwell in the echoing old chemistry building at the University of Leeds where I was preparing a lecture as a PhD student. I could try and make it sound romantic but the reality was that I was on the way to the toilet and I could smell the toilets from where I stood, staring at the wall as I thought. Like most good ideas, it was very simple, and it is highly likely that the same thought had already occurred to many others, but being an academic, I got to be the first person to write it down. I was editing the proofs of a paper later that week and slipped the idea into the introduction and waited to see what happened. The paper was published in *Science of the Total Environment* in 2003 and led to an article in the *Guardian* newspaper, suggesting that the restoration of peat bogs could be financed via carbon markets. This led to a call from an NGO which wanted to use my idea in their work, and I had to tell them that it really was just an idea, and that it would be at least 10 years before it could actually be used. But if they were up for it, I told them, I'd be happy to work with them to see if we could turn the idea into reality.

A full 12 years later, the UK government launched the policy instrument that had started life in that smelly stairwell in Leeds. I say 'launched' like there was a big fanfare or something. The reality was that we were given a microphone and a box to stand on during a drinks reception and half the audience spoke over the announcement. But it was official, and as a result of a huge number of people's work, building on that germ of an idea, we are now working with business, NGOs, government and landowners to save some of the UK's most beautiful and yet undervalued landscapes.

Ever since my PhD, working with cattle herders in the Kalahari Desert, I've been interested in how people can work together more effectively to tackle shared goals. And so increasingly, alongside my environmental research, I have been researching how people co-produce and share new ideas. I've been trying to understand how we can combine insights from research with the lived experience of local people and politicians to make better decisions.

During this work, I've seen first-hand the thirst for new ideas among tribespeople who have lost their grazing lands to thorn bushes and their cattle to drought. I've felt the embarrassment of discovering I was a government 'go-to person' on a subject I thought I knew little about (it turned out I knew more than them and I was the only researcher they could approach who could explain the evidence to them in policy-friendly terms). And like most researchers, I've had acute bouts of 'imposter syndrome' when I've been asked for my opinion as an 'expert' (the worst were the times I had to present research to United Nations conferences, with rows of people at desks labelled with their country names — I'm still not 100% convinced that was actually me).

But what have the privileged few (who get these opportunities) got that the rest haven't? I am living proof that you don't need to be any more intelligent, lucky or good-looking than anyone else. I suffer from the same sorts of insecurities as every other hard-working researcher I know, and I've struggled all my life with self-confidence (I actually had a panic attack halfway through my first lecture). However, through the research I have done, and my lived experience of implementing the findings of my research, I believe that there are a small number of key principles that can enable anyone to significantly increase the impact of their research. In this chapter, I'm going to introduce you to these principles.

Five principles to underpin your impact: tips from researchers and the people they've worked with

I often hear great ideas and case studies about engaging with people who might be interested in or want to use our research. Those ideas have ranged from the obvious (I can't believe I didn't think of doing that already), to innovative, unusual ideas (that I'd love more people to hear about). Recently, I decided it was time to look at these ideas more systematically. I wanted to find out what researchers around the world were doing, so that the research community as a whole could start to learn from what works.

As a researcher, I know that many of you will want to read the original research, but I'm not going to bore you with the details here (you can pop over to my personal website to read the papers if you're suffering from insomnia: www.profmarkreed.com). However, it is worth saying that they are derived from interviews with researchers and knowledge brokers around the world, and the stakeholders and members of the public that they worked with to produce (or in some cases co-produce) research outcomes. My colleagues, Dr Ana Attlee, Professor Lindsay Stringer and Professor Ioan Fazey, and I were funded by the UK Research Councils to convene a group of international experts in the study and practice of research impact (named in the acknowledgements) to distil principles that could be used to train researchers from every academic discipline. You can see a full list of the original research publications that this book is based on under 'Further reading' in Part 4, as well as a list of some of my favourite books and articles on the topics covered in this book, which have inspired my thinking.

Figure 2 shows what emerged from our interviews with researchers and stakeholders. The themes around the outside give you a flavour of what the interviews told us, and you can see how we've grouped them under the five principles in Figure 3. There is some overlap between the principles and they are not meant to be implemented in any particular order, though some principles will underpin the application of others. Table 2 captures some of the ideas people shared with us during interviews. These suggestions are adapted from quotes, based on a qualitative analysis of what researchers and stakeholders told us worked most effectively in their work. Hopefully those tips give you a flavour of the principles that follow.

Figure 2: Themes relating to effective knowledge exchange (KE) and research impact that emerged from an analysis of interview transcripts by Reed et al. (2014), showing how these themes map onto five principles of effective impact

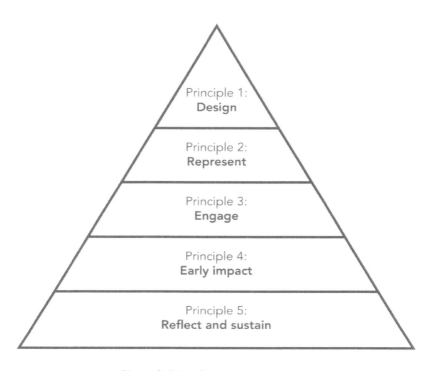

Figure 3: Principles to underpin impact

In the following chapters in Part 1 of the book, I'm going to unpack each of these principles briefly in turn, so that we've got a firm foundation upon which to start taking action and generating impact. I will illustrate each principle with practical suggestions about how you can apply them in your research. Part 2 of the book then builds on these principles to walk you through five steps to fast track your impact. Parts 3 and 4 then provide more detailed guidance on specific tools and templates you can use to facilitate impact in your research.

Table 2: Tips from researchers and stakeholders interviewed by Reed et al. (2014) about how to generate impacts from research.

Design
<u>Understand what everyone wants.</u> Spend time understanding what different stakeholders and project members want from the research. This can help with managing expectations and identifying potential issues/problems early on. It is important that all involved know the objectives of the project and what their role is likely to be, as well as the project outputs and any recognition they may gain from their involvement. Time spent at the beginning introducing each other and sharing personal motivations and goals is helpful.
<u>Understand the context of the project</u>. Understand local characteristics, traditions, norms and past experiences, and use this as a starting point for planning the project. Some projects have found it useful to carry out ethnographic research prior to starting their research to ensure their plans match the needs and preferences of local communities.
<u>Take your time.</u> Knowledge exchange is time-consuming if done properly. If not done properly, bridges can be burnt that will influence not only the effectiveness of the present project but projects to come. Plan for the time it will take to do knowledge exchange properly, including skilled staff time.
<u>Design your knowledge exchange activities carefully.</u> It is vital to plan the knowledge exchange process well. Spend time researching the context, the stakeholders and possible approaches. Look into alternative approaches, so you have a Plan B. Design for flexibility, get feedback and adapt your plans, and always try and adapt your plans to suit changing circumstances. It is best to plan to use a range of methods and approaches in the design of your knowledge exchange activities.
<u>The early bird catches the worm.</u> Knowledge exchange activities should be initiated early in the project. Ideally, planning and research into the context and stakeholders should begin prior to project commencement.

Get buy-in. Ownership and ongoing commitment to your research can be quickly established by getting 'buy in' from the key stakeholders. This can be formal (e.g. in the form of monetary investment or contracted time to the project) or informal (e.g. regular engagement via social media).

Independence. Ensure that the management of the research is seen as independent and neutral, so you can build trust with stakeholders. This can be achieved through a neutral organisation leading the process or an independent facilitator running sessions with stakeholders.

Mix up your methods. Plan to use a variety of methods for engaging with stakeholders and the public. Different people will enjoy and be best suited to different methods. Always start with those methods that will be the most comfortable for people, and as trust builds, more innovative methods can be used.

The process is as important as the outcome. How a knowledge exchange process is implemented is often as important as the final impact. Ensure proper attention is paid to creating an effective knowledge exchange process while keeping impact goals in sight.

Resource your impact. Generating impact takes significant time and resources. There are methods available for low budgets that are reliant on the team's personal time and energy. However, if budgets are too low, corners may be cut and outcomes may be compromised. Budget for a well-designed process, which includes social events, staff time, professional facilitation, refreshments and (in some cases) financial compensation to cover time and expenses incurred by those participating in your research.

Use knowledge brokers. Take time to identify individuals who play a significant role in your stakeholder community and who may be able to act as a champion for your work. Such individuals will be well known by many diverse groups, and able to understand their different perspectives. If you can build a strong relationship with someone like this, they can help you build trust with new groups by proxy.

Present your research visually and orally as well as in words. Aim to present information using sight and sound rather than in words where possible e.g. maps, illustrations, cartoons, drawings, photos, models and readings.

Represent

Involve the right people. Spend time researching which stakeholders are best to involve in your research. Make sure power dynamics between individuals are considered and attention is paid to selecting individuals who have the power to make a difference. Involve all parties as early as possible, preferably in the planning process. Time spent in one-to-one discussions to win over those who doubt the value of the process before you start is well worthwhile. If there are people or groups who cannot be convinced at the outset, keep them informed and give them the option of joining in later. Where possible, work individually with people who are particularly disruptive, to avoid disrupting group events.

Not just the usual suspects. Those of different ages, gender, backgrounds and cultures bring different knowledge, concerns and perspectives to the table. By representing the full diversity of interest in a well-designed process, a project can have a far greater long-term reach and sustainability.

Understand and create networks. Understand the social networks that the people you want to work with are part of. Spend time creating connections, both vertically and horizontally, within and between organisations relevant to your research, to ensure you have access to people with decision-making power and a resilient network of people engaged in your work.

Personal initiative. Many impacts from research are based on one individual's initiative, perseverance and hard work. To achieve impact, you need at least one individual who is willing to push the process through and maintain momentum.

Away days. Put time aside at the start of the project for the research team and key stakeholders to get to know one another's expertise, background and languages. Include time for socializing.

Be enthusiastic. Enthusiasm for your research and the process of engaging in your work is often infectious. Enthusiasm can help maintain momentum and achieve long-term involvement of participants, even when outcomes are delayed or mistakes are made.

Find out what motivates people. People are motivated to become involved in research for a number of reasons, for instance: academic interest, to learn, fear of missing out, financial gain, professional duty, personal promotion, and to support or promote causes they care about. It is important to take steps to make personal agendas explicit, perhaps through anonymous ballot at the start of the project or explicit discussion. If you are unlikely to be able to deliver what people want, make this clear from the outset. Be honest with participants about what they will gain through participation. Do not have a hidden agenda.

Build capacity for engagement. Create a shared skill base in your team for impact, and include basic training activities in the project early on to improve knowledge exchange and co-production.

Build personal relationships. Impact is all about relationships. Taking time to socialise is just as important early on in a process as time spent on knowledge exchange activities. Schedule in social time in the project and get to know participants on a one-to-one basis.

Build trust. A lack of trust can significantly hinder attempts to generate impact. Spend time explicitly considering levels of trust in the project and how to improve trust between team members and stakeholders.

Multiple modes of two-way communication. Whether face-to-face or via social media, use the widest possible spectrum of communication media available to you, so that everyone who is interested in your research can engage with you via their preferred mode.

Keep in people's comfort zones. Be aware of what is comfortable for those involved and keep within their comfort zone. Have meetings in the local area and in a non-threatening, neutral environment. Choose activities (at least initially) that people are comfortable with.

Enjoy! Make sure the process is enjoyable and interesting for yourself, your research team and everyone else involved.

Keep it simple. Do not assume certain levels of literacy or education. Keep language and approaches simple and accessible. Spend time discussing and agreeing terms to be used, and the best approach to take. A stakeholder steering group may help in ensuring the language and approach is suitable.

Work around people's commitments. Keep people involved by respecting and working around people's commitments. Consult with those you want to work with as soon as possible to match your process to their commitments. For example, it might work best to have morning meetings rather than evening meetings and certain times of year may not work well for the attendance of certain groups.

Manage power dynamics. Power dynamics can have a significant impact on your work with stakeholders and the public. It is incredibly important to recognise that power dynamics play a role in the process and to plan for and manage this appropriately. For example, ensuring a first-name basis can go some way towards balancing power but it is still important to recognise that others will be conscious of who holds a formal role in a hierarchy and will be adapting their behaviour and communication as a result.

Record. In order to ensure transparent, trustworthy processes, make sure that your process is properly recorded. This is also important in order to identify and learn from methods that have been particularly successful or unsuccessful. However, do be aware of methods of documentation, some participants may be uncomfortable with audio or video recording.

Keep your goals in mind. Reiterate research and impact goals throughout the process and keep to deadlines.

Respect cultural context. Make sure that your approach is suitable for the cultural context in which you are working. Consider local attitudes to gender, informal livelihoods, social groupings, speaking out in public and so on.

Respect local knowledge. All participants will have significant knowledge of their community and will be capable of analysing and assessing their personal situation, often better than trained professionals. Respect local perceptions, choices, and abilities and involve all types of knowledge when setting goals and planning for impact.

Share responsibilities. Share out responsibilities and credit in order to help build relationships, trust in the process and foster ownership for those involved.

Early Impacts

Deliver quick wins. Ensure that if the project aims to create practical outcomes it delivers on these. Delivery of practical outcomes is a key motivator for involvement, and identifying 'quick wins' for delivery early on can help build trust and relationships, keeping people engaged so you can deliver longer-term impacts

Work for mutual benefit. Work hard to ensure that the project is of mutual benefit. Spend time finding out what people want from the process and try hard to deliver this. Unbalanced processes, for example, those which appear to be all about academic benefit, can fail to get the best from those involved and can affect trust and commitment to the process.

Reflect and sustain

Get participant feedback regularly. Ensure that you get feedback throughout your research on how your activities are being perceived and participants' concerns/ideas. Such feedback will help the project to adapt techniques and deal with problems as they arise to improve the effectiveness of impact.

Make time for reflection. Build in time for all involved in impact generation activities to reflect on the process and outcomes. This is especially important when working in areas of conflict to ensure optimum learning and behaviour change.

Learn from others who have achieved impact. Spend time exploring similar work and institutes within your area. Visit other projects that successfully delivered impact and speak to people who have carried out similar work to what you are planning. It may be useful to engage a mentor from a project you admire and ask them to give feedback on your process as you go along.

Continuity of involvement. Continuity of people involved is important, especially for projects dealing with some form of controversy. By including the same group of individuals, critical relationships and trust develop, which facilitates impact.

Maintain momentum. Regularly monitor progress to ensure that initiatives are built on and objectives achieved or altered as required. Impact takes time and often takes unpredictable turns. If there has to be a break, start from where you left off and build this into the process. It may be useful to call a break for a period of reflection and present it as part of the process. Review sessions, feedback forms and good facilitation can ensure that momentum is maintained.

Chapter 4
Principle 1: Design

Summary

Know the impacts you want to achieve and design impact into your research from the start:

- *Set impact goals from the outset*
- *Make an impact plan*
- *Build in flexibility to your plans so they can respond to changing user needs and priorities*
- *Find skilled people (and where possible financial resources) to support your impact*

Set goals and plan for impact

I think we are all pretty good at coming up with research questions and setting objectives for the new knowledge and insights we want to derive from our research. However, most of us have little experience of developing impact goals or identifying objectives relating to the knowledge exchange activities we will use to achieve those impacts. Whether you're at the start of your PhD project or initiating a multi-million dollar research consortium, now is the time to set goals for your impact. Even if you are halfway through your project, it is never too late to make a plan. Although you won't be able to do all the things you would have been able to do if you'd planned for impact from the beginning, it is often surprising how many opportunities there still are to change your approach to the project so that you can extract as much impact as possible from your work.

Many research funders now ask us to identify these goals as part of the application process. Increasingly, having a strong impact plan (or 'pathway to impact' as it is sometimes called) is essential to get funding (whether this is a condition of funding or it just makes you more competitive). However, it is surprising how many researchers leave this part of their funding application to the last minute. Planning for impact at the last minute means you don't have time to discuss your plans with the people who might benefit from your work, you won't be able to evidence any form of partnership or co-development of your impact plan (whether in your text or via letters of support) and so may not be very believable, and you will miss opportunities for impact that would have been easy to identify had you left yourself more time.

It is also surprising how few researchers revisit that part of their application when they get their funding and initiate the research. Just because you got the funding doesn't necessarily mean that your plans for impact were particularly good. Your plans may have passed a quality threshold, but the bar is sometimes set quite low. So, it is always worth revisiting your plans to make them more detailed and actionable. You will need to start planning now for the activities that you will have to undertake alongside your research, if you are to have a chance of realising your impacts. You need a plan.

In Chapter 10 in Section 2 of the book, I'll help you create your own impact plan. You can also see detailed guides to writing the impact sections of proposals to UK and EU funders in the resources section of my website, and you can look at examples of good practice at: http://www.fasttrackimpact.com/pathways-to-impact. First, however, I would like to explore why planning for impact is so important, and how it can help you save time and become more efficient in your pursuit of impact.

Set goals and plan for impact

I think we are all pretty good at coming up with research questions and setting objectives for the new knowledge and insights we want to derive from our research. However, most of us have little experience of developing impact goals or identifying objectives relating to the knowledge exchange activities we will use to achieve those impacts. Whether you're at the start of your PhD project or initiating a multi-million dollar research consortium, now is the time to set goals for your impact. Even if you are halfway through your project, it is never too late to make a plan. Although you won't be able to do all the things you would have been able to do if you'd planned for impact from the beginning, it is often surprising how many opportunities there still are to change your approach to the project so that you can extract as much impact as possible from your work.

Many research funders now ask us to identify these goals as part of the application process. Increasingly, having a strong impact plan (or 'pathway to impact' as it is sometimes called) is essential to get funding (whether this is a condition of funding or it just makes you more competitive). However, it is surprising how many researchers leave this part of their funding application to the last minute. Planning for impact at the last minute means you don't have time to discuss your plans with the people who might benefit from your work, you won't be able to evidence any form of partnership or co-development of your impact plan (whether in your text or via letters of support) and so may not be very believable, and you will miss opportunities for impact that would have been easy to identify had you left yourself more time.

It is also surprising how few researchers revisit that part of their application when they get their funding and initiate the research. Just because you got the funding doesn't necessarily mean that your plans for impact were particularly good. Your plans may have passed a quality threshold, but the bar is sometimes set quite low. So, it is always worth revisiting your plans to make them more detailed and actionable. You will need to start planning now for the activities that you will have to undertake alongside your research, if you are to have a chance of realising your impacts. You need a plan.

In Chapter 10 in Section 2 of the book, I'll help you create your own impact plan. You can also see detailed guides to writing the impact sections of proposals to UK and EU funders in the resources section of my website, and you can look at examples of good practice at: http://www.fasttrackimpact.com/pathways-to-impact. First, however, I would like to explore why planning for impact is so important, and how it can help you save time and become more efficient in your pursuit of impact.

Collaborative planning for impact

The best impact plans are collaborative (at the least) or co-produced with those who will benefit from the research (at best). You can only get so far planning impact with your research team. Unless you are really embedded within the groups of people who will use your research, it will be difficult to really understand what motivates them, and what their needs and priorities are. If you can develop your impact plan with these people, then you are far more likely to engender a sense of shared ownership over the work and genuinely meet needs. When people feel invested in your research, and know that they will benefit tangibly from your success, they are far more likely to stick with you when things are taking longer than expected. They are far more likely to be forgiving when things don't go according to plan. And they are far more likely to help you achieve your impact. Apart from giving you their time, many organisations will actually give you access to resources and staff, which can greatly enhance your capacity for impact. From your perspective, you have willing and enthusiastic partners who can significantly increase the chances that your research will have impact. From their perspective, they're getting access to top researchers and the credibility of your university brand at a much lower cost than if they were paying consultants to help them with the same tasks. In one of my research projects, I asked stakeholders what motivated them to work with researchers, and the findings may be useful to bear in mind if you want to tap into the motives of the people who are most likely to use your work (Box 1).

Developing an impact plan in collaboration with the people who are interested in your research doesn't come without risks. You need to be careful to avoid raising false expectations, as you will not have the time, resources or expertise to do everything that people suggest. If you've done your stakeholder analysis right (see the next principle and Chapters 9 and 14), then you shouldn't discover yourself in a room full of nurses if you're researching archaeology.

However, you may need to extend your research to include some of the things people want to see. In my experience, publics and stakeholders will very rarely object to you asking certain research questions, but they may ask you to extend your research to consider additional things you hadn't previously considered. Not everything will be possible, but it is often surprising what can be achieved with some lateral thinking and collaboration. If it truly is unachievable within the context of your project, then see if there is someone else you could put people in touch with, or if there is funding you could apply for that could enable you to do what they would like to see.

The extent to which you plan for engagement with stakeholders and the public will differ from project to project, depending on how applied or close to impact it is, or how easy it is to communicate with public audiences. For projects that are not yet very applied, where you know there will need to be significant investment of time and resources to transform the work into something that is interesting or useful, it is appropriate to plan for less engagement at the outset. It is not reasonable to expect someone to spend a significant amount of time working with you throughout a project from which there is not likely to be any tangible benefit to them. But even if you think there's another 10 years of research and development ahead of you, it is always worth engaging early with the potential users of your research. Although the company or organisation may not exist by the time you're actually ready to achieve your impact, they may be able to connect you to others who can help you and identify alternative or interim uses for your work that you wouldn't have thought of yourself.

Depending on the nature of your work, it is often worth taking advice to ensure you protect your intellectual property before talking to potential commercial partners. However, rather than letting the fear of being ripped off stop you talking to people, try and work out how you can reach out to potential users and start talking about your ideas safely, so you can make the connections (both socially and intellectually) that you will need to keep moving forwards towards your impact.

Box 1: Top reasons why people work with researchers

When we asked stakeholders what motivated them to work with researchers, these were the most common reasons people cited:

- Accessing future funding and new business opportunities
- Developing new solutions to old problems
- Increasing personal impact/influence through collaboration with researchers
- Intrinsic motivation to "make the world a better place" or a desire to learn about the issues being researched

Bear these motives in mind, and see if you can work out which of these motives apply to the people you want to work with. By tapping into their motives, and explaining clearly how working with you can achieve what they want, you are much more likely to get the level of engagement you want.

Plan the timing of your engagement

For projects that are likely to yield useful or interesting results in the near future (even if not during the lifetime of the project), it is important to offer as many opportunities as possible for stakeholder and public engagement throughout the research cycle. Where possible, engage these groups during proposal development and research planning (e.g. using 'seed-corn' funding sources). It is always possible to spot funding applications that have been developed in this way. They have powerful letters of support (that are actually written by the person who signs them, rather than the research team) and their plans for impact are detailed and credible, with named contacts who clearly already trust the research team. In many fields, it is also possible to engage these groups in data collection, analysis and/or interpretation of results, for example, prioritising, ranking and evaluating your findings.

For research on contentious issues, co-production of research in this sort of way can have advantages and drawbacks. On one hand, if done effectively, co-production of research that engages stakeholders from opposing sides of a conflict can act as a powerful means of achieving a form of consensus that no amount of additional evidence on its own could have achieved. It becomes hard for groups who don't like the findings to criticise the research when they were involved in its design and execution. On the other hand, if you fail to reach consensus or do not involve certain powerful groups, the legitimacy of your research may be undermined by claims of bias (based purely on the people who were involved in the research).

In some of my projects, I will engage the two opposing sides of the debate I'm researching to make them aware of the beneficial impacts of the research going 'their way' and prepare them for the eventuality that the research findings will 'go against' them. If it becomes clear that the research findings are not what one side of the debate want to hear, then you are able to hear their objections to the work before it is in the public domain, and if these are valid objections, you may be able to refine the research to take these factors into account before finalising the work. If the research findings remain the same despite taking into account the various confounding factors that your opponents would have used to undermine or discredit the research, then these groups will find it much harder to undermine the research if they know you took their concerns seriously and have already addressed them. Although this is unlikely to make them any happier with your research findings, they will typically be grateful to have had the chance to input to the work and at least have advance warning of research that might damage their interests, so they can prepare for this. As a result, you retain trust with both sides of the debate and can continue working together in future, without compromising the rigour or independence of your work.

If you have engaged with key groups throughout the research cycle, then it becomes easy to work together to communicate findings at the end of the project. Each of the organisations you have worked with is likely to want to put out its own press release, and will want to promote the research through their networks. The challenge at this point becomes co-ordinating dissemination activities (but your press office will usually be happy to do this for you). For more specialist communication to specific groups or key decision-makers, it can be invaluable to work with stakeholders to design your communication.

Working together, you are likely to be able to find the key words and framings that are most likely to make the messages from your research resonate with the right people. And where necessary, these organisations may be able to take your message to these people, whether by opening doors, accompanying you or actually being the messenger. This can be important for communicating findings to key decision-makers who would otherwise be out of your reach, for example, a government minister.

Decide what to do if you think your research might make money

Discovering that your research has commercial value is a mixed blessing for many researchers. Many researchers who come up with valuable intellectual property (IP) as a result of their work feel overwhelmed by the options they are presented with. Not wanting to abandon academia for business, many researchers do very little with the opportunities that arise, and as a result squander the chance to generate impacts from their work. However, you don't have to change careers to exploit your IP. There are three ways to get as much impact as possible from your ideas, without having to give up your day job.

There are several ways to commercialise your ideas or IP to realise impact from research. Each one has it up- and downsides and the decision to go down one path over another will depend on the nature of your ideas, your institution, your personality and the level of commitment you are comfortable with. In all cases you should talk to your Technology Transfer Office (TTO) to assess the commercial opportunities of your IP and look at the potential commercialisation mechanisms available to you. This will involve looking at potential applications and users of your IP, developing a commercialisation plan and taking some initial steps.

The three typical routes available are:

1. **Make your ideas available for others to use on licence:** many academics worry that if they negotiate with companies, their best ideas may get bought up and then shelved by corporations that want to protect their existing products from your new ideas. The good news is that you don't have to give up the rights to use your work if you grant them a **non-exclusive licence**. With this arrangement, you can licence your IP to several different companies, and increase the likelihood that one of them manages to bring your ideas to market. Be careful though; if a company asks you to **assign** your IP to them, talk to your TTO and look at your options, or you may lose all rights over your IP. Your TTO will deal with the contracts. Then a third party pursues your impact, with minimal effort from you. Easy.

2. **Target a dream partner who can take your ideas to scale:** this option takes a little bit more work, but not a lot. The problem of licencing your ideas to anyone who is willing to pay for the

privilege (the non-exclusive licence in the point above) is that these companies may not have the capability to develop your ideas to their potential. Even if they are capable of developing your ideas, they may have a different timescale to you and have quite different ideas about the sorts of impacts they want to see from your IP. Instead, consider working with your TTO to find a dream partner who shares your passion and priorities, and has the capabilities to develop your ideas in ways that will give you the impacts you most want to see. If you want to retain the right to develop your IP yourself in future, you need to go for a **sole licence**. A sole licence enables you to target a single organisation that you would like to develop your ideas, and you both have the right to use the IP. An exclusive licence allows the licensee to develop the IP in the confidence that no one else can access the IP. This can strengthen a research collaboration with your dream partner but caution is required. You will need to ensure you have chosen the most capable organisation to realise the potential of the IP.

3. **Spin-out companies** are the hardest work, but if you are prepared to invest some time in the beginning, your TTO can help you put a team in place to run the business for you, so you don't have to quit the day job. You may be able to set up as a not-for-profit company enabling you to invest profits in charitable work linked to your research. The spin-out route gives you the greatest amount of control over realising the impacts you want to see from your research. However, many businesses need to raise capital before they can launch, which may expose your personal finances to risk if you take loans, or compromise your control over the company if you bring in investors. If you go down this route, you will need to work closely with your TTO to develop your ideas, as your institution will have a vested interest.

However busy you are, it is worth spending time negotiating the right deal if you want to commercialise your IP. Third parties and potential investors may have quite different motivations to you, so it is essential to have a really clear vision of the impacts you want to see arising from commercialisation before you go into any negotiation with commercial partners. Flexibility is necessary in any negotiation, but your TTO can help you get a deal that works for business, impact and your time.

Resource your plan

Part of the impact planning process I will go through in Chapters 10 is about making sure you properly resource your plans for impact. To an extent, this is simply about planning the knowledge exchange activities you want to undertake in detail, so that you can budget for them properly. But it is also about the staff resources you have to support impact. Although most research organisations now employ staff to support impact, it is rare that such staff can be deployed to specific projects, so you will need to make sure you have team members in place with the right skills. For larger projects, this may take the form of a paid position on the team.

For smaller projects (including PhD projects), this isn't possible, but you can still get help. It is often surprising how far a small budget can go if you can find a reasonable designer or consultant who has a track record of working with researchers (for example, take a look at Fast Track Impact's Design for Impact service). If you are really working on a small budget, then it is worth looking at what skills there are around you. Reach out to others in your research group or graduate school. Offer what you're good at, and find out what others have got to offer. Perhaps you are a native English speaker and can help people proofread or write for generalist audiences, and there may be others in your group who are great at photography or have designed their own website.

Stay flexible

A good impact plan is not set in stone. Public interest in your field may suddenly peak as a result of a related issue in the news, and stakeholder needs and priorities change as they adapt to dynamic business and policy environments. So, it is useful if you can build an element of flexibility into the design of your research. This is really important because often, once your research design has been funded, you are expected to deliver it as it was designed. Building an element of flexibility into the design will help you justify changes you need to make as the research progresses, to accommodate the changing needs and priorities of those you are working with.

This is easier for some types of research than others. I was always told that reviewers would not look favourably on grant proposals that incorporated flexibility, but framed effectively I have found that a certain amount of flexibility actually attracts praise from reviewers. You can still set ambitious goals and promise clear outcomes, but by incorporating flexibility you can credibly claim to be able to adapt the research to the needs of those who will benefit, so they benefit more. For example, I often include an exploratory work package in which additional research questions and outcomes can be defined by stakeholders, and I will budget for events and dissemination materials that I do not yet know the focus of. If your plan for impact is flexible, you can be opportunistic and exploit opportunities to communicate your research and generate impact as they arise. If you're locked into a plan that is too rigid to change, with no support or resources available when you need them, opportunities may pass you by.

Be strategic to save time

Once you have an impact plan, you are able to see that there are multiple groups of publics and/or stakeholders, who will be easier or harder to reach. You will be able to see the different pathways to impact mapped out in front of you. Some will be quick, easy and likely to work, while others may be more expensive, time-consuming and uncertain. This knowledge empowers you to choose which groups you want to work with first, and which pathways to impact you will prioritise based on the time and resources you have available. Rather than trying to achieve your dream list of impacts for every group that has a stake in your work, you may prioritise particular impacts that are significant and far-reaching, but that are quick and easy, and tried and tested. If your chosen pathway to impact doesn't work, you may want to prioritise some of the other pathways you mapped out as a plan B. And if your circumstances change and you end up having time, resources or expertise, you can move onto the other impacts and groups.

Chapter 5
Principle 2: Represent

Summary

Systematically represent the needs and priorities of those who will use your research:

- *Systematically identify individuals, groups, organisations and publics that are likely to be interested in, use or benefit from your research*
- *Identify stakeholders who could help or block you, or who might be disadvantaged by your work*
- *Revisit who you're working with as your context and stakeholder/public needs and interests change*
- *Embed key stakeholders in your research*
- *Consider the ethical implications of engaging with different stakeholders at different stages of the research cycle*

As researchers, we are increasingly expected by funders to identify and incorporate 'beneficiaries' into our work from the outset. Working out who might benefit from our work isn't always easy though. Even if we know who will benefit from our research, an equally important but often unasked question is:

"Who might be disadvantaged or lose out as a result of my research?"

Even if we can answer both of these questions, there is another crucial question that every researcher should ask themselves:

"Who has the power to enable me to do my research and achieve impacts, and who has the power to block my work?"

It is just as important to identify individuals, organisations, groups and publics who might be disadvantaged by the outcomes of our work, or who may block our research, as it is to know who our beneficiaries are, and who can help us. I'm going to call these people 'stakeholders' and 'publics', rather than 'beneficiaries' (see Box 2 for my definition of stakeholders and publics). Knowing about potentially problematic or hard-to-reach stakeholders or publics at the outset can give us the necessary time to adapt our research so that it no longer disadvantages those groups, or work out ways of mitigating negative impacts before we run into opposition or achieve bittersweet impacts for one group at the expense of another.

I was once approached by a researcher who told me that I was her last hope, and that if I couldn't help, her next call was going to be to her funder to give them their money back. She was doing research on the police, but it turned out that someone at the top of the police force was ideologically opposed to what she was doing and so blocked her from getting access to any policy force in the country. She couldn't do her research, let alone achieve the impacts she wanted. One of my pathways to impact was delayed by years after I learned that a major environmental non-governmental organisation (NGO) was ideologically opposed to my research, and would actively block my colleagues and I from achieving the policy impacts we wanted to pursue. Rather than try and oppose this powerful organisation, I formed an alliance with another environmental NGO which was interested in our research, and through them, we worked on changing the views of key people in the opposing organisation. Only once their board had taken the decision not to oppose our pathway to impact (they still would not support us) did we then start talking to the policy community. It took another 12 years for that NGO to eventually make a decision at board level to support us in our pathway to impact. By then we had already achieved significant policy change. To this day, I doubt that we would have achieved what we did if we had started our pathway to impact without realising we had such a powerful opponent.

Analysing publics and stakeholders

It may seem self-evident that all the relevant publics and stakeholders should be identified prior to any attempt to engage. However, it is surprising how often this step is omitted in research projects that need to work with non-academic partners. In many cases, this omission can significantly compromise the success of the research. For example, the project may miss crucial information that could have been provided had they engaged with the right people.

In cases where very few stakeholders are identified or engaged with, this can lead to a lack of ownership of project goals, which can sometimes turn into opposition from certain stakeholders. In cases where a single important stakeholder has been omitted from the process, that organisation or group may challenge the legitimacy of the work, and undermine the credibility of the wider project. Publics/stakeholder analysis helps solve these problems by:

1. Identifying who has a stake/interest in your work
2. Categorising and prioritising stakeholders/publics you need to invest most time with
3. Identifying (and preparing you for) relationships between different stakeholders and publics (whether conflicts or alliances).

I will explain in more detail how to perform a publics/stakeholder analysis in Chapter 14. At this point, I want to explain the benefits of a publics/stakeholder analysis:

1. **Start talking early to the right people**, so that you can identify any major barriers to your work, and identify the people who can help you overcome those barriers. There is evidence that projects that engage with stakeholders/publics early engender a greater sense of ownership amongst people who are then more likely to engage throughout the lifetime of the project, and implement or use the outcomes of the work you have done together.

2. **Know who you need to talk to**: don't just open your address book or talk to the 'usual suspects'. Find out who might lose out, as well as who will benefit. Identify who is typically marginalised and left out, as well as the people and organisations that everyone knows and trusts. For example, you might draw on methods from the arts to identify stakeholders using tacit knowledge or past experience. Those who are left out are usually the first to question and criticise work that they feel no ownership over.

3. **Know what they're interested in**: you need to have a clear idea of the research issue at stake before you will be able to effectively identify stakeholders/publics. But that doesn't mean that the research questions and issues you explore together should be set in stone. As you begin to identify stakeholders/publics, you will find out more about the nature of their stake/interest in your research, and you may need to broaden your view of what is included in your work, if everyone is to feel that their interests are included. Be aware that some organisations may have very different interests in your work, depending on which team you speak to. Also consider that most groups will have many different interests, but you are principally concerned with the interests that intersect with your research, rather than their wider interests.

4. **Find out who is likely to benefit most from your research:** some publics will benefit more than others by virtue of their specific interest, needs or characteristics. For example, if you are researching young people affected by knife crime, young people from certain backgrounds living in certain areas are more likely to benefit from your research than other publics. Some organisations might have specific needs that your research can meet, which would provide them with much more

significant benefits than other organisations which may be larger and better known, but which do not have needs your research can meet.

5. **Find out who's got the most influence** to help or hinder your work; some people, organisations or groups are more powerful than others. If there are highly influential stakeholders who are opposed to your project, then you need to know who they are so that you can develop an influencing strategy to win their support. If they support your work, then it is also important to know who these groups are so you can join forces with them to work more effectively. There will be some influential groups who have relatively little interest in your work. For example, they may have a broad remit that includes many issues that are more important and urgent to them than the specific focus of your research. Influential individuals are often busy and inaccessible, and you may need to spend significant amounts of time and energy getting their attention before you are able to access their help.

6. **Find out who is disempowered and marginalised**: publics/stakeholder analysis is often used to prioritise more influential groups for engagement. Although time and resources may be limited, it is important not to use this analysis as a tool to further marginalise groups that are already disempowered and ignored. Many of these groups may have a significant interest in your research, but very little influence over the issues you are researching, and little capacity to help you achieve the impacts you want. However, there may be compelling ethical reasons for engaging with them.

7. **Identify key relationships so you avoid exacerbating conflicts and can create alliances** that empower marginalised groups. It can be incredibly valuable to know in advance about conflicts between individuals, organisations or groups, so that you can avoid inflaming conflict and, where possible, resolve disputes. Through publics/stakeholder analysis it can sometimes become possible to create alliances between disempowered groups and those with more power, who share similar interests and goals, thereby empowering previously marginalised groups.

For more detailed guidance on how to do publics/stakeholder analysis, see Chapter 14.

Box 2: What are stakeholders and publics?

A stakeholder is any person, organisation or group that is affected by or can affect a decision, action or issue. Rather than just identifying 'beneficiaries', a publics/stakeholder analysis seeks to identify people, organisations or groups who may be either positively or negatively affected by your research. In addition to identifying those affected by your research, publics/stakeholder analysis seeks to also identify those who might affect your ability to complete your research and generate impacts, either positively or negatively. These stakeholders might not directly benefit from or be negatively affected by your work, but they might have the power to enable it or prevent it from making a difference. Publics/stakeholder analysis is covered in more detail in Chapter 14.

Although everyone may be considered a member of the public in certain contexts, it is important to recognise that there are differences between individuals by which we can group them, e.g. backgrounds, affiliations, gender etc. Rather than thinking of the public as a single entity, it is useful to start thinking about different 'publics' if we want to identify groups who are more likely to be interested in our research. By targeting engagement activities towards these specific publics, it is possible to engage more efficiently and meaningfully. Chapter 14 explains how you can use a publics/stakeholder analysis to identify different publics, and introduce more advanced methods for segmenting publics.

Prioritise who you contact to save time

Thinking systematically about who you engage with can save you significant amounts of time. Many researchers engage reactively with whoever reaches out to them, whether or not these people will benefit strongly. As a result, they spend large amounts of time working with groups that will benefit in small ways, when they could have spent a much smaller amount of time reaching out proactively to a smaller number of groups that would have benefited much more significantly. I am not suggesting we should turn people away if they are looking for help, but taking a more strategic approach to your engagement using something like a publics/stakeholder analysis can enable you to justify spending less time with certain groups, or signposting them to opportunities to learn about your work that are more efficient (such as a large group event, public lecture or newsletter).

A publics/stakeholder analysis will give you a list of the most important groups you need to reach out to. You decide on what basis they are important. For example, some people focus first on those with greatest interest who will benefit most or have most influence. I tend to prioritise hard-to-reach groups that would benefit significantly from my work but are not that interested, because I know it will take more time and effort to work out how to engage them than other groups. Now that you know which groups are most important for you to proactively reach out to, it is possible to be much more strategic with your time. You may only have 15 minutes per month to dedicate to impact, but if you use that time to write one email to the top three groups on your list, three months later you will be significantly closer to impact.

Chapter 6
Principle 3: Engage

Summary

Build long-term, two-way, trusting relationships with those who will use your research and co-generate new knowledge together:

- *Have two-way dialogue as equals with likely users of your research*
- *Build long-term relationships with the users of your research*
- *Work with knowledge brokers and professional facilitators*
- *Understand what will motivate research users to get involved*
- *Work with stakeholders to interpret findings and co-design communication products*

This principle sits centrally within between the five principles, and it is key to any attempt to generate impact. True engagement is about deeply understanding and empathising with the people you are engaging with. In fact, if you wanted to boil my whole approach to impact down to a single word, it would be empathy.

Engage with empathy

At the heart of this principle is the idea of really connecting with the people who might be interested in your research. Empathy is putting yourself in someone else's shoes. It is understanding what it is like to think and feel like someone else. Or, as Barak Obama put it:

"The ability to put ourselves in someone else's shoes; to see the world through the eyes of those who are different from us — the child who's hungry, the steelworker who's been laid off, the family who lost the entire life they built together when the storm came to town. When you think like this, when you choose to broaden your ambit of concern and empathise with the plight of others, whether

they are close friends or distant strangers; it becomes harder not to act; harder not to help."

If you can sense what it is like to be that other person, then you will know what motivates and inspires them, what worries and challenges them, and you will be able to find ways of making your research relevant to their needs and interests. No amount of tools and techniques can substitute for this skill. To paraphrase the apostle Paul, it doesn't matter how persuasively you craft your message, if you are unable to empathise with the people you're communicating with, then all they'll hear is noise.

Some people believe that empathy is something you are born with. You either have it or you don't. I believe that empathy is a skill that can be learned. It is harder for some and comes naturally to others, but it is something we can learn, practise and get better at. One of the key tools you can use to put yourself in the shoes of your publics or stakeholders is the publics/stakeholder analysis template I described in the previous chapter.

Empathy is particularly important in situations where you need to deal with multiple competing agendas and personalities. The more ownership people feel over your research, the more likely it is that they may try and exert influence over the research process and outcomes. Where these suggestions might compromise the rigour of our research, it is clear that we should not follow them. However, many suggestions could take us down equally legitimate and interesting routes, and by following these paths we may be able to effect significant change that benefits many. The problem is that different stakeholders and publics may have competing visions of the path they would like the project to take, for example, focusing on different case studies or focal issues. Add to this long-standing conflicts between different individuals and organisations and you've got a challenging meeting on your hands. Negotiating between these competing agendas will be tricky, and will require every bit of emotional intelligence you can summon up.

Sometimes a connection happens over the research, and sometimes it might be over a picture on their wall or a book on their shelf. The point is that I'm trying to understand something about what really makes them tick; what gets them up in the morning and keeps them going. If it's not obvious, then I will ask people indirectly at first, with questions like, "What took you into this role? What is it that you love most about what you do? What is it about this organisation or task (or whatever it is that they are doing) that you love?". Then, more directly, I may ask what sort of unanswered questions (relating to my field of enquiry) they have, what their goals are, and what knowledge or help they need to reach those goals, individually or as an organisation. Usually there is something somewhere in the answers to these questions that enables me to connect them to my research or a colleague, and something that enables me to connect with them at some slightly deeper level too.

Many research programmes now offer formal opportunities for researchers and stakeholders to work more closely together, for example, via work shadowing, placements and fellowships. These are invaluable opportunities for understanding stakeholder organisations and their cultures, and for demonstrating the benefits that your research can bring them. If there aren't any formal opportunities you can apply for, then create your own opportunities to connect with people through the events you attend, and the way you design events (e.g. providing sufficiently long breaks, designing activities to get different people working in small groups together, or opportunities to talk *en route* to a site visit).

Engage with humility

The second thing that I think we can do to cultivate empathy is to be humble in our interactions with stakeholders and publics. It is remarkably easy to fall into the trap of living up to people's expectations, without even realising that you're doing it. If someone expects you to act like the know-it-all expert who will give them all the answers, social conditioning creates a strong subconscious desire to meet that expectation (especially when doing so is likely to give you an ego boost at the same time). Resist that temptation. A simple expression of humility, for example, telling someone that you don't know the answer to their question, and that you would genuinely like to hear their opinion, can instantly break down barriers, and build trust and rapport. When it becomes clear from your questions and your behaviour that you genuinely want your research to help them, and it's not just about building your career and ego, people start to tentatively offer small things that you might be able to help with. No matter how small that first request is, if you can do everything in your power to help, and go the extra mile, you will gain enough trust to be asked to help with something bigger.

Engage in two-way dialogue

There is an under-explored set of values that are implicit in any attempt to generate impact *with* rather than *for* people. Genuine two-way dialogue requires listening, and listening requires mutual respect. To learn from what you hear requires humility, especially when what you hear challenges your assumptions. From a place of humility, we no longer regard ourselves as the sole authority on the matter, and are open to the possibility that we might be able to learn something from people with very different backgrounds and types of knowledge to us. If we see the world in black and white terms, where the role of researchers is to discover universal truth, then there is little point in engaging with other researchers in debate, let alone stakeholders and publics.

If, on the other hand, we see the world in shades of grey then it becomes possible to accept that there may be multiple options rather than a single solution, and multiple ways of seeing things rather than a single preferred interpretation. Decisions in this 'shades of grey' world are as likely to be informed by lines or argument and moral reasoning as they are by formal evidence from research. Taking this approach, listening to perspectives from stakeholders and publics becomes a worthwhile endeavour, and we become able to engage in meaningful dialogue, rather than just communicating or disseminating our research.

In 2011, Jeremy Philipson and colleagues (see 'Further reading') reviewed seven commonly used knowledge exchange mechanisms and found that participating in events was most likely to be associated with benefits for policy and practice (91% of research projects reported impacts arising from event participation). However, these benefits were typically one-way, with limited two-way benefits reported for the research. In contrast, they found that engaging more deeply with stakeholders as project partners or via advisory panels was more likely to generate win-wins for both researchers and stakeholders. Projects engaging with stakeholders as project partners or via advisory panels were perceived to improve research quality (100% and 87% of researchers perceived a positive effect from project partners and advisory panels respectively) and relevance (100% and 95% respectively), as well as impacting positively on stakeholder knowledge (81% and 93% respectively), and policies or practices (77% and 64% respectively). Two-way engagement works. To find out how to set up your own advisory panel for a research project, see my guide in Part 4 of the book.

Engage for the long term and build trust

One of the most common and inadvertent ways that researchers break trust with their stakeholders and publics is when they break contact at the end of a research project. Researchers typically work on 3–5 year project cycles, and during a project may work intensively with specific groups or organisations. However, once the project is over there are often few reasons to maintain engagement. As a result, when the researcher comes back to an organisation after a number of years to request a letter of support and engagement in a new project, they may be met with scepticism. If you missed a few important emails from the organisation during that time, you may further compound the perception that you are only interested in engaging with them to get research funding and further your career, not because you genuinely care about their interests.

As a result, it is worth thinking about ways of keeping contacts 'warm' between projects in a time-efficient manner when you have no project staff to support you. Make sure staff who work for you on the project introduce you to the contacts made so you don't lose touch with them when your project staff move on. Consider writing a newsletter or setting up a social media channel to provide people with news of your work and other material between projects that may be of interest to them, so you retain open channels of communication in ways that do not take too much time.

It is these long-term, trusting and two-way relationships that foster knowledge exchange and ultimately lead to impact. People talk of creating 'safe spaces' for collaboration with stakeholders, but no matter how well you design your workshop, you can offer little safety as a stranger. It is those trusting relationships that actually persuade people to come to the workshop in the first place, and it is those trusting relationships that become the life-blood of your impact between events and other interactions.

Engage a facilitator when you can

If you can afford it, professional facilitation is one of the most important factors leading to successful outcomes from workshops with stakeholders (Box 3). A professional facilitator can help create spaces for constructive dialogue and collaboration between research teams and stakeholders, especially in situations where there is conflict or controversy. Despite having undertaken professional training as a facilitator, for high-stakes, high-conflict situations or contexts where I am not independent and so it would not be appropriate to facilitate, I hire a professional facilitator. For the most important workshops and meetings I will hire the best facilitator I know and pay thousands, but you can get more affordable facilitators, especially if you are able to design the session yourself.

If you can't afford professional facilitation, there are still a lot of simple things you can do to design meetings and events that effectively facilitate themselves. Chapters 15 and 16 provide a range of tools and techniques that you can use to design events with stakeholders so that you can easily keep a handle on the power dynamics and prevent things getting out of control. Broadly speaking, I progress through the following types of techniques when I'm designing a workshop (for detailed examples of methods in each category, see Chapter 15):

- **Opening out**: There are a number of techniques for opening up dialogue and gathering information with stakeholders about issues linked to your research. This collection of techniques is particularly useful during the initial phases of a research project, either during the development of initial research questions prior to writing a funding proposal, or in the early phases of a funded project, where the research goals and programme of work are being adapted to better fit the needs and interests of stakeholders.
- **Exploring**: There are a range of methods that can help you evaluate and analyse preliminary findings with stakeholders. Given the length of most research projects, getting early feedback on preliminary findings can help keep stakeholders interested in the process and give them greater ownership over the eventual research outcomes. The feedback can also provide researchers with ideas about how to further refine their

work, for example, where assumptions are not clear or are questioned by stakeholders.

- **Deciding**: After issues have been opened out and analysed, it is often necessary to start closing down options and deciding on actions based on research findings. There are a number of techniques that can engage researchers and stakeholders in decisions based on research findings, for example, prioritising particularly interesting or relevant findings for further research or action.

Box 3: Three secrets of successful stakeholder engagement

As part of the British Academy-funded Involved project, I investigated 24 projects in 20 different countries in which researchers worked with stakeholders in an attempt to work out the most important factors that enabled those processes to achieve their goals. The three most important points that emerged from the analysis were:

1. Represent all the relevant stakeholders;
2. Get a professional facilitator to help you manage power dynamics between stakeholders; and
3. Empower stakeholders with information and decision-making power, so they can meaningfully participate in your research.

The groups that got these three things right were not only more likely to achieve the goals they had established together, but they were also more likely to report having learned and gained trust in the other participants as a result of their experience. The networks and alliances that can arise from such interaction may yield impacts many years after your initial work together.

I also believe that a few tips as a facilitator can go a long way if you don't have a lot of experience. Although there is no substitute for experience, having a plan B (and C) and a few ways of dealing with challenging individuals can turn a scary situation into a productive and enjoyable event for everyone. Chapter 16 provides lots of practical advice on facilitating workshops with stakeholders.

Engage to co-design your communications

Working with stakeholders to identify and articulate the implications of research for policy and practice can help researchers target their communication effectively and enhance the probability that the target audience interprets findings appropriately. Co-designing communication materials with stakeholders can increase the likelihood that other stakeholders engage with the material, and the process of co-developing materials may facilitate learning, both among stakeholders (about the research) and researchers (about how to communicate more effectively with particular groups). If it is possible to get research users and stakeholders to disseminate these communication materials themselves, this may further increase their reach.

Save time by working with knowledge brokers

You would be right to think that there is no substitute for time when it comes to building two-way, trusting, long-term relationships. However, there is one crucial way you can save significant amounts of time if you want to engage effectively with stakeholders or publics. Knowledge brokers work like bridges between disparate social networks and can significantly speed up the process of accessing particular groups or organisations, and gaining trust. Effective knowledge brokers are typically well known and trusted by many different groups, and have an interest in your research. Because they are known and trusted by many of the stakeholders you want to work with, if you are introduced or recommended by this intermediary, people are much more likely to trust you.

There are many academics who act as knowledge brokers. These individuals don't typically advertise themselves as such, but if you can work out who they are, they are usually very approachable and often keen to help. For example, I act as a knowledge broker through my role as Research Lead for the International Union for the

Conservation of Nature's UK Peatland Programme and can connect researchers with key charities, government departments and practitioners that might otherwise be hard to access. Put out some feelers in your department, university or disciplinary networks and you may be surprised at who might be able to short-cut you to relationships with key people who can help you achieve impact. It may be worth creating official roles for such individuals in the research, such as including them in advisory panels or involving them in hosting or co-designing events.

Chapter 7
Principle 4: Early Impacts

Summary

Identify quick wins where tangible impacts can be delivered early in the research process to reward and keep likely users of research engaged with it. However, be aware of the ethical implications of engaging people early in the research cycle.

One of the key challenges of achieving impact from research is the timescale on which most research projects operate. At best, most research projects take three years to complete, but some last longer, and it is often closer to five years before findings are actually published in the peer-reviewed literature.

In contrast to this, most businesses, practitioners and policy-makers need knowledge to address specific needs on timescales measured in days, weeks or months. Many of the staff who embark on the project with us at the start are no longer in post by the time we have findings to share with them. As a result, we can create the impression that we are more interested in what they can give to us to help our research than we are in helping them.

If you are working with members of the public, providing benefits early in your research can enable them to shape it and stay involved throughout the course of your project, rather than just at the end. Early engagement can get people interested in your research and build contacts and mailing lists you can use later on to engage these same groups with your findings.

Manage expectations about timelines

The first challenge, then, is to manage expectations about the timescales on which you will be able to deliver findings, making it clear that you will not be able to deliver *final* results till the end of the project, which is typically years later. Once you have overcome this hurdle, the next challenge is to retain people's interest and engagement, as you get on with the hard work of doing your research. I've been involved in projects where we engaged with people at the start to set the agenda and then didn't get back in touch again till we'd finished the research. For us, every day brought a new challenge, and we were so busy that it felt like very little time had passed before we contacted people again. In contrast, people reacted with surprise that we were still there, still doing the research — many had assumed that we'd forgotten all about them and their concerns. In the intervening period, many of them had moved on — to other jobs and other concerns. Had we stayed in touch, the people who had moved into their roles would have known about our research, and we'd have had a chance to adapt our research to their changing needs and preferences.

Creative ways to produce quick wins

So, how can you deliver tangible outcomes for people early in a project, before you've actually done the research? I'm not suggesting that you compromise the rigour of your research by leaking early findings to the press before you've gone through the proper ethical and peer-review processes. I think that there are a number of creative ways you can answer this question without compromising your ethics or the quality of your work that can help keep people motivated and engaged in your research (Box 4).

I think the first answer to this question is to recognise that people value research for its ability to provide them with answers to questions; knowledge that is new *to them*, relevant, interesting and useful. As researchers, we are so focused on the generation of completely new knowledge that we forget that our knowledge of the discipline already enables us to answer many of the questions that people care most about. I think we often forget the privileged position we sit in, behind journal pay walls, with an understanding of the jargon that enables us to access the latest knowledge in our field as it is published. Many of the people who most need this knowledge are unable to justify paying to access the research, and even when they do, they find the language impenetrable.

Just by making our own existing knowledge accessible (let alone the latest knowledge in our field), we can immediately add value. For example:

- **Run workshops, exhibitions, performances or other events about the general subject area of your research** (e.g. an author or composer's work, or hearing perceptions) to build connections and mailing lists of relevant groups who are likely to be interested in your work, once research findings become available (e.g. about a particular aspect of that author's work, or your new audiology research).
- **Crowdfund a product or performance related to your research in your first year**. The crowdfunding will give you the resources to run this as a mini project in parallel with your main research, and you then have a diverse and interested audience

who you can contact with future updates and events related to your research (see my guide to crowdfunding research in Section 4).

- **Make your review of the 'state of the art' from your funding proposal available as a briefing note with links to the latest literature.** Most of us have already reviewed the latest literature as part of our research proposal before we even started doing the work. Can you make this short review available in accessible language to people who are interested in your work? You can even do this from failed research proposals — the research you proposed may not have been good enough to get funded, but there was probably nothing wrong with your review of the literature in the field (depending on how original your perspective is on the literature, you can even turn this into a 'response' to one of the key articles you've reviewed, and contribute to the peer-reviewed literature — I've done this).

- **Turn the literature review you did at the start of the project into a briefing note.** Most research projects start by reviewing the literature in greater depth, but these reviews end up buried in PhD theses or are condensed to form a section of a paper that only appears years later in an inaccessible journal. Can you turn your literature review into a briefing note for stakeholders or send it to a freelance journalist to see if they can help you turn it into a feature piece for a weekend newspaper or magazine? I once got into some trouble with a piece of government-funded research I was advising on, where the commissioning officer was upset at the lack of progress on the project. As far as the team were concerned, they were just getting on with the research, but there was nothing to show for their work yet. I had reviewed the literature in order to be able to advise the team, but the review I'd sent them was highly technical. So I suggested that I turn my review into a couple of policy briefs. I didn't have time to do any design or polishing, but that evening I created two briefs and by the next day the policy team were loving the project and already moving forward in their thinking about the issue. If you're not sure how to write a policy brief, go to Chapter 21.

- **Write a newsletter**, and don't worry if there's no news to report from your project. Doing a monthly or quarterly survey of the latest news and peer-reviewed literature relating to your project is a good habit to get into anyway. The chances are that you're already getting all this information delivered to your email inbox and social media accounts, and incorporating that knowledge into your mental model of the issues you're researching. Creating lay summaries of key papers is a great way of honing your generalist communication skills, and it often surprises me how much this helps me understand and remember a paper I've read. Once you've done this, you can instantly share the key insights in a form that's easy to understand with a link to the original paper on social media. You can also harvest news stories from social media that others may not have come across and link to these from your newsletter. I've published project newsletters during quiet periods where we're just getting on with the research, which have very little project-specific news but which retain interest and create value for stakeholders by reporting the latest news in our field. Although this takes time, in addition to adding value for your stakeholders you'll be surprised how much value it adds to your research if you have to always be on the cutting edge of what's happening in your field.

- **Create a powerful and useful social media presence for your research**. Consider how useful your social media presence is for those interested in your research. If you already have a well-focused group of people enjoying what you do and you want to reach out more widely, consider creating new social media accounts that specifically target particular groups or reach out to the public and promote these from your main accounts. I once created a Twitter account for a research project that became one of the most influential in its field, and was eventually rebranded and taken over by an international NGO after the project ended. Of course, I've also created accounts that were never really noticed by anyone. But with a clear social media strategy, it is surprising how much value you can add by promoting research and evidence in an easily accessible way. You can find out about how to create a social media strategy for your research in Chapter 17.

- **Reframe your research as press releases to link to the latest headlines and topical issues.** If you are aware of the latest issues hitting the news, you will be well placed to adapt or reframe your research, so that you can put out a press release that links to a particularly topical issue which is far more likely to get your work media coverage. It was this sort of media coverage (linking peat bogs to issues of climate change) that initiated the impact I described at the start of Chapter 3.

- **Co-ordinate milestones** in your research with the milestones that your stakeholders are working towards. If you know about a forthcoming policy review, election or issue-based campaign, product launch or event in the diary of those you're working with, it is sometimes possible to reorganise your research to provide relevant findings in time. Moving your data collection forward by a month may be relatively easy for your team, and could make the difference between being able to contribute to the development of a new policy or product, or not. If you don't know about these milestones till they are almost on top of you, there is often very little you can do to provide meaningful contributions in time.
- Depending on your field, and the ethics procedures you need to go through, it can sometimes be possible to **make data and models available to stakeholders** to analyse and test themselves. For example, you may not have time to produce anything meaningful before a deadline in the policy world, but policy analysts may be very grateful to have access to data that they can run their own analysis on to brief their minister before a debate.

Adapt to changing needs and priorities

It is important to regularly discuss your work with stakeholders if you want to make sure you are producing work that will still deliver relevant impacts, and to spot opportunities to add value as you conduct your work. These people will quickly tell you if their agendas have moved on and they are now looking for different things from your work — if you ask them. They will also be able to tell you how to make your work as accessible as possible, so as many people as possible can benefit from it.

For example, in one project I was managing we were planning to use Geographical Information Systems to show people how a landscape was likely to change in the future. We thought that this would be really accessible because it visualised some fairly complex model outputs in pretty colours on a map. However, when we took this idea to our stakeholder advisory panel, they told us that they found this really abstract and difficult to get their heads around. Instead they asked us why we couldn't just make a documentary film to explain it. This took some additional funding, which we were able to find by juggling our project budget a bit, but it was actually a very easy request to meet. Those films have now been watched thousands of times on YouTube, giving us far more reach than we could have had using maps in workshops. This illustrates the first principle (design, Chapter 4), which emphasises the importance of building flexibility into your project and having stakeholders and publics embedded in your work.

Consider the ethical implications of your quick wins

Finally, it is important to stress under this principle that I'm not advocating unofficial publication of results prior to proper quality control. Many researchers have been tempted to release findings early when their topic has hit the headlines, and regret their actions when they later discover flaws in their work, which would have been ironed out if they had waited. Sometimes you have to let opportunities go, and trust that other opportunities will arise once you are confident that your findings are ready to be made public. In cases where you think it is acceptable to provide preliminary findings or data to stakeholders for their own internal uses, it is important to consider carefully how you communicate risk and uncertainty relating to your work, to avoid misleading people. In some cases, if you are taking a co-productive approach to your work, key stakeholders will be embedded within your team, with access to data. What happens to your data then comes down to trust, which emphasises the importance of the previous principle (engage, Chapter 6).

One project I was leading was co-funded by both the government and an NGO that was opposing the government on the issue we were researching. I received some very worried emails from civil servants when it became clear that our findings contradicted government policy ahead of a major policy announcement. At the same time, I received emails from the NGO saying that they thought they could get front-page news headlines based on our research, which would help them achieve their goals. There was sufficient trust in the team, however, that I was able to speak to the head of the NGO and suggest that they would be more likely to change government policy if they worked constructively with the civil servants who had jointly commissioned the work because they wanted to base their work on the best possible evidence. The policy announcement went ahead (but with the caveat that it would be revisited and possibly changed in future) and was denounced by the NGO without reference to our research. Then, over the following year, the NGO and others worked with government using our evidence and other research, and the policy ended up being changed.

In another project, someone created a company to exploit our research at a fairly early stage. This was a quick win for us so we worked with the company, but it became clear that they didn't fully understand the research, and we started hearing stories that the Chief Executive Officer (CEO) was having meetings with other companies, trying to sell a product that didn't actually exist. I tried speaking to the CEO, who assured me that he understood the science and would stick to it from now on without making false claims or promises. However, he then went on Sky News and said a lot of things that were factually incorrect and misleading. I called up the chair of his board of directors to express my concerns, and he called me back the following week to say that the board had met, sacked the CEO and dissolved the company. The next company that came along to try and exploit this same piece of research was led by someone who had collaborated on parts of the research, and they are now successfully selling a product that actually does exist.

Engaging with publics and stakeholders throughout the research cycle can bring up a range of ethical dilemmas. The stories I have shared with you from my own experience are just a small flavour of the challenges you may face. In some rare cases you may be able to predict ethical challenges when you are designing your pathway to impact. For example, with more experience I might have predicted that taking funding for an applied and policy-relevant piece of research from the government and an NGO that was challenging the government on its policy in this area would lead to problems. In the second example, it wasn't possible to predict the challenges that lay ahead, though I have to admit that alarm bells were ringing the day I first met the CEO of the company exploiting our work. I couldn't put my finger on what it was, but something about him screamed out 'used car salesman'.

So, have a think about how you can provide early impacts from your research — but not too early. You may be surprised how much value you can add already.

Chapter 8
Principle 5: Reflect and Sustain

Summary

Keep track of your progress towards impact, so you can improve your knowledge exchange, and continue nurturing relationships and generating impacts in the long term:

- *Track your impacts*
- *Regularly reflect on your knowledge exchange with your research team and stakeholders*
- *Learn from peers and share good practice*
- *Identify what knowledge exchange needs to continue after the end of the project and consider how to generate long-term impacts*

How do you know your research is having an impact, and who knows about your impact? Although these questions seem simple to answer, researchers and institutions around the world are scratching their heads, trying to work out how to actually monitor the impacts they are having on the world around them.

Reflect on both your knowledge exchange activities and impact

The last principle is all about becoming a reflective practitioner of knowledge exchange and impact. I can't think of a single example of research impact that did not involve some sort of knowledge exchange. Therefore, as I said in Chapter 3, if you want to have an impact, you need to be great at knowledge exchange. If you want to be great at knowledge exchange, then you need to take time to reflect on your activities and get feedback. Evaluations of impact are also often required by funders and governments who want to see the impact that their money is having. The same information can also be used for marketing purposes, and to target resources and support for the researchers and teams who need it most. By sharing good practice across an institution, it becomes possible to celebrate individual and collective success and learn, so that everyone can work together to make a bigger difference.

There are two broad types of approach to evaluating impact: summative and formative (Box 5). Summative feedback evaluates the extent to which you achieved impact. Formative feedback evaluates whether your pathway to impact is working. This formative feedback is essential to enable you to improve your knowledge exchange practice, and keep impacts on track.

Box 5: Two approaches to evaluating impact

There are two broad types of approach to evaluating impact:

- Summative evaluation and evaluation of impacts after they have occurred with minimal participation of researchers or beneficiaries, to provide ex-post measures of reach and significance
- Formative evaluation and evaluation of knowledge exchange and impact in collaboration with researchers and beneficiaries, to provide ongoing feedback on reach and significance, so that impacts can be enhanced during the research cycle

The majority of impact evaluation is done in the first, summative, ex-post mode by research funders to evaluate the impact of their investments or to distribute quality-rated funding to the best research institutions. However, it is also worth investing in formative tracking of impacts as they arise, including an ongoing evaluation of the knowledge exchange activities that are meant to deliver those impacts. This will let you know if activities are not working as intended and can provide an early warning that your pathway to impact is off course. This information can be crucial in enabling you to get your impacts back on track.

Do I really need to evaluate my impact?

There are many different reasons why researchers monitor and evaluate the impact of their work. As individual researchers, many of us are motivated extrinsically to demonstrate impact so that we can get on in our career. We may be intrinsically motivated to see if we are achieving our personal goals, so we can learn how to make more of a difference. Research and impact managers in our institutions may want us to track impacts so they can understand how to target help to those of us who have the potential to make a difference.

Much of this is extrinsically motivated by the need to achieve high scores in impact evaluations that will feed into rankings, reputation and financial rewards for the institution. Although much of this might feel like game-playing, and probably is, I have yet to meet a knowledge exchange professional in an impact-orientated role who gets their kicks from playing the game. Ultimately, our goal collectively is to help each other raise our game to make our research more relevant and change more lives for the better. If we can change the narrative from bean-counting to progress-watching, then more of us might be persuaded that impact can go far beyond promotions and rankings. Evaluating impact might become more about sharing good practice and inspiring each other to go the extra mile. After all, the impacts we leave from our research are a huge part of our legacy. As Jackie Robinson, the first African American to play Major League Baseball, put it: *"A life is not important except in the impact it has on other people's lives."*

The challenges you will need to overcome

The challenges of evaluating and then evidencing impact are not trivial. There are particular problems with attributing impacts to specific research findings, which are exacerbated by time lags between the production of knowledge through research and its application and impact. There are also challenges associated with how impact is defined and perceived, both by researchers and those we work with. There are concerns that current attempts by research funders and governments to measure and reward impact may be funnelling effort into a narrow range of socially constructed impacts that are considered by researchers to really matter.

I was recently part of a team that reviewed 135 evaluations of knowledge exchange and impact across diverse disciplines. We found that researchers in different disciplines defined impact quite differently, and hence looked for quite different things in their attempts to track it. Partly this is simply due to the fact that some disciplines (e.g. many from the arts and humanities) focus more on public engagement and education, while others focus on more instrumental impacts. However, we found that one of the most important things determining people's definition of impact and how they measured it was their perception of what constitutes valid knowledge (what social scientists refer to as 'epistemology'). Those with a more 'positivist' perspective tended to focus on one-way knowledge transfer to achieve impacts, and were more likely to track impact using quantitative methods. Those with more 'subjectivist' perspectives, in contrast, were more likely to run knowledge exchange activities that encouraged mutual learning through multi-stakeholder interactions. People who viewed knowledge in this way were more likely to track impact using methods that captured the diverse experiences of those involved and were more likely to consider a wider range of factors that may have contributed to impacts.

I think this calls for some fairly deep reflection, which may not be comfortable for all of us. How do you think knowledge is generated and what do you consider to be valid knowledge from which impacts should arise? What, then, do you think are the most relevant ways of tracking your impact?

Sources of feedback

Broadly speaking, there are two main sources of formative feedback if you want to reflect upon and improve your knowledge exchange practice. First, feedback from colleagues can be incredibly helpful — these are the people who you can (hopefully) trust to be constructively honest with you when things really didn't work, and who can support you to improve your practice. I like to schedule time for a debrief with my research team after any kind of knowledge exchange activity, even if this is an online activity, so we can discuss what went well and what could have been done better.

Second, some of the most valuable feedback can come from the members of the public and stakeholders we were trying to engage with. Often, a survey at or after an event can give you the feedback you require, but sometimes you just need to pick up the phone or meet up for a drink with someone to find out in depth what went wrong or right. If you wait till the end of your project to ask people, it may be too late to correct simple issues (like the time of day you were engaging with people) that could have made a massive difference to the levels of engagement with your research.

As discussed in the previous principle (early impacts, Chapter 7), checking in regularly with stakeholders can also help you identify new opportunities for generating impact that you would not have spotted if you hadn't sought their feedback. Also, by building in opportunities for regular feedback about your knowledge exchange activities throughout the project, it may be possible to reassess who holds a stake or is likely to use the research, to ensure that stakeholder representation remains relevant throughout the research cycle (represent, Chapter 5).

Where possible, involve stakeholders who have engaged with the research in the evaluation of your knowledge exchange activities. The evaluation process provides opportunities for stakeholders to work together to share perspectives, increase ownership of and responsibility for knowledge exchange, and enables participants to work together to refine your practice. The challenge is to develop evaluation processes that facilitate achievement of desired goals, provide valid and reliable data for ongoing planning and decision-making processes, and contribute to open and accountable research governance. I will discuss methods for evaluating and evidencing impact in greater depth in Chapter 22.

Find ways to keep track of your impacts efficiently

If you've developed an impact plan for your research (see the first principle, engage, in Chapter 6 and the template in Part 4 of this book), then you should have some simple metrics you can monitor to keep track of your impact. I will discuss how you can evidence your impact in more detail in Chapter 22. However, given the time lags and other factors that can affect your impact, it is just as important to track the success of your knowledge exchange (see the indicator column in the impact planning tool in Part 4). If you monitor your knowledge exchange, you can improve your practice, and enhance the likelihood that your research actually does lead to impacts.

This sounds great in theory, but the problem is that very few researchers actually do this. The number one reason for this is that researchers feel they don't have time to record and track their impacts. Many universities now have online repositories for researchers to report their impacts, but they typically get limited engagement from researchers. Ask yourself why you don't regularly or systematically keep track of your knowledge exchange and impacts. Based on the answer to this question, try and develop a system that meets your needs so you actually do keep track of things. Some people keep ring binders and scrapbooks, others create a folder in their email system, and still others use cloud-based systems. I personally use the productivity app Evernote as a way of storing information about impacts that I can use in seconds, offline, on a mobile device, without having to remember another username and password (those are the four reasons why I don't use my institutional repository regularly and the excuses I used to use to avoid keeping track of my impacts). Now, when I'm asked to enter my impacts into someone's system or report to my funders, I can look in one place and pull out the most relevant material quickly and easily. You can find more on this in Chapter 22.

Make legacy arrangements for your impact

Finally, as part of your reflection, it is worth considering what might happen to your impacts if you leave your post, or finish your PhD or move to a new country or discipline. Substantial benefits can be derived for many stakeholders after projects have been completed, through ongoing communication and interpretation of findings. Where possible, 'legacy arrangements' can support continued engagement between researchers and research users, to extract and augment value from the previous research through interpretation activities and supplementary analysis. It can help if the length of time over which engagement needs to be sustained is considered from the outset. For example, if a project plans to develop a network that will have the potential to work together beyond the time frame of the initial project, it will be necessary to forge collaborations with organisations who share this goal, but who can also fund or administer such a network long after the project has ended.

Despite wanting to move my research from Africa to the UK after my PhD to avoid becoming an absentee father, I maintained relationships with those I worked with and got funding to follow up my research through the production of manuals in local languages for agricultural advisors and farmers. This has led to a steady flow of opportunities to work with the policy community, which in turn have spawned a steady flow of academic publications. Partly because of the expertise that these publications demonstrated, but I suspect partly also because of my reputation for public speaking, this led to invitations to speak at two major United Nations policy conferences. What has surprised me is how little effort it took to keep these relationships alive, and add value, despite this no longer being a major focus of my research. People are thirsty for knowledge, and hungry for people who can give it to them in palatable forms.

Part 2: Steps to fast track your impact

Chapter 9
Step 1: Envision your impact

Summary

This first step will help you fast track your impact by understanding the sorts of impacts that your research might be able to generate. You will need to come up with a core goal and understand the motives behind why you want your research to have more of an impact. Finally, you will break this ultimate goal down into a series of smaller steps that you can take over the coming weeks and months in your research.

Five steps to fast track your impact

Understanding each of the five principles in the preceding chapters is essential if you want to take a relational approach to impact that has far-reaching and long-lasting effects. Although some of the principles are a little abstract, I've tried to illustrate each of them with practical things I've done, and which you can also do to enact each principle in your own research. But actually taking these principles and putting them into action can be hard without a bit more structure. So I've developed five steps based on these principles that can enable you to change the way you do research, so you achieve impact.

This will work best if you have a research project in mind that you'd like to derive impacts from. I recommend that you take about a week to think through each step and try out the tasks in your own research. This gives you the opportunity to learn about impact in the specific context of your own research discipline and gain confidence as you try things out in the real world, one step at a time. These steps are based on evidence I've collected over my decade of research on impact. They are also based on my own personal experience, and the experience of the many researchers

I've worked with to take their impact to new levels. They really do work — try them out.

If you want extra help with these steps, you can sign up to receive each step via email every week over five weeks on the Fast Track Impact website (http://www.fasttrackimpact.com/for-researchers). I'll talk you through each step in a short video, and I'll give you personal help with your ideas for impact.

These are the steps (Figure 4):

1. **Envision** your impact
2. **Plan** for impact
3. **Cut back** anything hindering or distracting you from your impact
4. **Get specific** about the impacts you will seek and the people who can help you achieve impact this month
5. **Achieve** your first step towards impact and monitor your success

This chapter introduces the first step, and the subsequent four short chapters explain each of the other steps. First, let me outline the steps and explain how they build on each of the principles we've looked at so far.

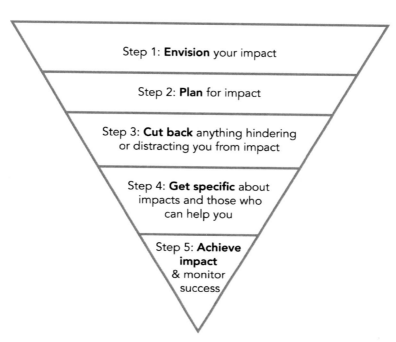

Figure 4: Steps to fast track your impact

Step 1: Envision your impact

Understanding the impacts of our research is obvious for some, but it is incredibly tricky for many of us. By the end of this step, you will have a clear idea of the types of impacts that your research might be able to generate.

I'll start by walking you through a series of questions designed to draw out the potential impacts of your work. Then we'll begin to focus on your core goal and start to interrogate some of the motives behind this goal. And finally, we will break this ultimate goal of your research into less intimidating, more immediate staging posts that can keep you motivated and on track.

Ten questions to identify your impacts

So, what impact do you think your research might be able to generate? If you are working on applied, real-world issues in your research, this question might be obvious. Sadly, for most of us, this is far from being the case. I've therefore developed ten questions to try and walk you through a process that should start to identify the potential impacts that could arise from your research:

1. Other than researchers, who might be interested in some aspect of your work?
2. What are they interested in, why are they interested and how might they benefit as a result of engaging with your work? If you can't answer this, go and speak to one of these people and ask them why they are interested and how they benefit from this interest.
3. What aspects of your research might be useful to someone, or could you (or someone else) build upon parts of your work to create something useful at some point in the future?
4. Going beyond your research for a moment, think of issues, policy areas, sectors of the economy, practices, behaviours, trends, etc. that link in some way to your research. What problems or needs are there in these places, and what are the barriers that are preventing these issues from being resolved? Could your research help address these needs and barriers in some way?
5. What is the most significant area of current policy, practice or business that your research might change or disrupt?

6. Which individuals, groups or organisations might be interested in this aspect of your research (whether now or in the future)?
7. What aspects of your research are they likely to be most interested in, and what would need to happen for this to become more relevant to them? What could you do differently to make your work more relevant to these people? Who would you need help from?
8. If these people took an interest in or used your research, what would change? How would you know they had benefited? What specific things would you notice or be able to measure? In the future, what about your research might people say was transformative for them?
9. Might you see changes in individuals, groups or organisations, or at a societal or some other level?
10. Would these changes be beneficial or might some groups be disadvantaged in some way as a result of your research?

I'll discuss how you can avoid negative impacts in the next step. So for now, please list as many benefits as possible for each of the different groups you identified in question 6. I'll show you how you can do this more systematically using publics/stakeholder analysis in Chapter 14, but at this point I want you to quickly get a feel for the people who might be able to benefit from your research.

Getting clear on your goal

At this point, you should have a number of different impact goals. Now, if you could only achieve one of these goals, which would it be? Which one is most important to you?

There are no rights or wrongs here. A goal may be important to you for purely personal reasons. For example, you might prioritise an impact goal that is likely to help you score highly when your research gets assessed, and will help you get a promotion. Or you might want to choose the goal that you believe will make the biggest difference to the issues and people you care most about.

The first thing that we are trying to do here is to get really specific and focused about what it is that we want to achieve. It is often surprising how opportunities suddenly appear that help you achieve a goal when you bring that goal very clearly into focus. This is probably due to nothing more than the fact that we notice things that can help us achieve that goal, which we would have otherwise missed. When you have a clear vision of where you are going, it is possible to cut out much of the noise that has been holding us back and creating confusion. We start to see clearly what it is that we need to do.

Ask yourself what your number one impact priority is. If by the end of your career you had only achieved one single non-academic impact from your work, what would you want it to be? There will be other goals, but by singling out the one that is your highest priority you are inadvertently flushing out some of the deeper motives behind your pursuit of impact, which will enable you to harness a deep source of motivation to achieve your aspirations for impact.

To make this easier, you can go through the benefits you have listed in response to the questions in the previous section, grouping similar benefits together and turning them into impact goals. Ask yourself "what is the good I can do?" or "what are the benefits I can provide?". If you are struggling with this, close your eyes and imagine yourself a few years in the future, looking at the most inspiring impact you can imagine from your research. What can you see? If there is a person standing in front of you, what are they saying about the value or meaning of your work to them? Make your picture as detailed as possible in your mind's eye, looking for evidence that you made a difference. You should now have everything you need to make a SMART impact goal: specific, measurable, achievable, realistic and time-bound. To explain what I mean by SMART in this context, contrast these two impact objectives:

- To curate a highly successful exhibition of work discovered as a result of my research (not very SMART)
- To curate and successfully market an exhibition of work discovered as a result of my research that will attract in excess of 10,000 visitors to the gallery by the end of the exhibition, and that will reach audiences across the UK and in at least three other countries via coverage in mass media and articles in specialist magazines (measured in column inches) and internationally via social media and online content production (measured via engagement metrics), leading to increased footfall and revenue to the gallery and increased awareness and changes in attitudes among audiences linked to the research (SMART)

What is motivating you to have an impact?

The second thing I want you to do is to work out what is really important to you personally about generating impact. The chances are that if you really analyse why it is that you have chosen this particular impact goal, you will find that it links in some way to your personal priorities and values, or to your identity as a researcher. Perhaps you need to get that promotion so you can move to a larger house because you want to have children, and it is those family values that are driving your desire to engage with the impact agenda. Perhaps it is simply ego; you want a legacy you can be proud of. Increasingly, many of us are motivated by institutional imperatives to generate impact, which in turn generate funding, improve rankings and benefit our employers. If these are the sorts of values that are motivating your engagement with impact, then you need to pay very careful attention to the identification of risks in the next step.

The reality is that all of us have complex and mixed motives for most of the things we do, but if personal benefit is a big motivator there is a danger that you may end up inadvertently creating negative unintended consequences in your attempts to generate impact. You would probably never think this consciously (although I was once in a stakeholder workshop with a senior academic who actually said this out loud), but you may be 'using' stakeholders to achieve your own goals, and as a result, stakeholders may well feel used.

Of course, our motives are usually mixed in everything we do, and ego and other personal benefits are usually part of that mix. We need to get a healthy balance of motivations driving our desire to generate impact, and make sure that we are not only or primarily being driven by extrinsic, instrumental motives. Being clear about our motives can really strengthen our motivation to generate impact. Bringing the complexity of our motives to consciousness can also help us to avoid some of the abuses of impact that many commentators have warned of if academics become more extrinsically incentivised to generate impact. And by focusing clearly on the myriad reasons why we are doing what we are doing, we are more likely to be able to pick ourselves up again when things go wrong, as they inevitably will at some point. If you want to think more about this, I've explored the power of our (often hidden) motives to empower or demotivate researchers in my book *The Productive Researcher.*

So, before you move to the next section, list some of the reasons why you want to generate impact, and try and put them in ranking order, with your highest priorities at the top of the list.

Where can you create most value?

I personally believe that rather than focusing on what we can get personally from engaging with impact, it is much more empowering to ask what we can give that produces benefit. Rather than remaining in a place of insufficiency, by focusing on what we do not have, this second question comes from a place of abundance and strength, which builds your confidence and empowers you to make a difference. I believe that generally the people who have achieved most in academia have been those who have given the most.

Their secret is strategic giving; they give to others who, like them, also give, rather than to those who consistently only take. That's not to say we should not mentor and help those who cannot give anything to us in return, that we should marginalise those less fortunate than ourselves, or that we should be calculating the time we give in terms of a return on investment. It does, however, mean that you might want to take a strategic look at whether some of the relationships you are investing most time in for your work are actually taking you away from spending time achieving your core goals. There may be very little you can do at this point but, if nothing else, it may be a lesson for the future (e.g. not to take on weak PhD students who aren't working in areas that are central to your research interests).

If you want to be successful at achieving impact, then give to others. Seek out how you can add value and help, and do all this in an attitude of humility, being open to learn from those you work with.

What will you achieve in the next six months?

The problem with the sorts of goals that many of us are trying to reach to make an impact is that they are often a long way off; in some cases, a very, very long way off. We probably all have experience with long-term goals; they have a habit of not happening. Things change; we change. Other things become more important.

So in addition to whatever your ultimate impact goal might be, no matter how far off or challenging it may be to achieve, I want to bring your focus closer to home now. I would like you to try and come up with some specific goals that are no more than six months in the future, but which are still closely linked to your ultimate goal. These will be staging posts on your journey to impact, and they will keep you motivated and provide you with feedback, and so keep you on track. The more detail with which you can visualise these goals, the more useful these short-term goals are likely to be. If you can, imagine yourself having just reached that staging post and see what it looks and feels like in your mind's eye.

Coming up with short-term impact milestones

There are a number of ways you can come up with these short-term milestones on the way to your ultimate impact goals. I'm going to suggest two that have worked for me. First, you can 'backcast' from your ultimate goal. As the name suggests, this is the opposite of forecasting. Start with the impact you want to achieve from your research, and then use your imagination to think of the step that would come immediately before having reached that goal. In the same way, keep stepping back, till you get to smaller initial steps that you might be able to take in the next six months.

The second approach works the opposite way around. Instead of working back from the end point, you need to look at where you are now in relation to your goal. The first thing you are doing is to identify everything you can that you've already done that has put you in a position to be able to pursue this impact goal, for example, you've got a funded research project and a team with some useful skills. Maybe you already have some really relevant relationships and have a clear understanding of what research needs to be done before you could have an impact. These are all strong foundations upon which you can build, and rather than looking at all your weaknesses, problems and barriers to achieving impact, you are moving into a more empowering place that recognises the strengths and achievements you can build on.

Box 6: Building on your strengths

Answer these questions yourself or with someone you know well. It can work particularly well if you do this in a pair with someone you know well, and then swap answers, and discuss. You will often find that your partner is able to identify many additional strengths you may not have been aware of.

1. Describe one of your greatest successes
Describe a high-point experience during the last year when you felt most alive, engaged or really proud of yourself and your work.

2. What are your strengths and skills?
Thinking of this experience, or another recent achievement in your work, identify what strengths and skills you brought to this experience that enabled it to be such a success. Consider strengths and skills you use in your work, and outside your work, and whether or not you can see an obvious way yet that these skills could help you generate impact.

3. What do you value most?
What do you most value about yourself and your job? Rather than just adding to your list of strengths and skills, consider whether there are some more fundamental things about who you are as a person and the role you play in your work that drive your success. These could be aspects of your personality or deeply held values and beliefs about yourself or the world.

4. What strengths can you build on to achieve even greater success?
Drawing on all the strengths, skills, expertise and value you know you bring to your work, identify strengths you believe you can either i) enhance and make even stronger; or ii) complement with new skills and strengths in future. Then consider the actions you could take to enhance or build upon your strengths.

Build on your strengths and successes

From this position of strength, then, you can then begin to see all the skills, insights and resources at your disposal, with which you might be able to add value to someone else who is on a similar journey. It can be useful to actually create a list of your skills and strengths. In Box 6 you can find questions to help you identify and build on your strengths for impact.

Apart from being a useful exercise for building your CV, this can be an enriching and revealing process; especially if you do it with someone who knows you well. In addition to the things that come immediately to mind, for example, areas of expertise in which you are highly knowledgeable or methods and equipment you can use, this exercise will enable you to identify other things that might not be so obvious. For example, do you enjoy photography or art in your spare time? Might you be able to take some of that creative flair into your impact work to add value to someone else? When you start to really look for places where these skills can add value it is often surprising how many opportunities will present themselves to you. For example, there may be a charity or business working in a similar area with similar goals which could really benefit from your expertise. By adding value to others and giving in this targeted way, you open the door to powerful new relationships and collaborative possibilities that would otherwise not have been available.

If you are really struggling to come up with any sort of specific impact goal for your research, simply looking for ways to add value with what you have got can often be the first step towards finding a tangible impact. By forming relationships with others outside the academy who are working in related areas, we often make mental connections between our research and the context these organisations are working in, which would not have been possible if we had not reached out.

Your tasks for the first step

1. Work through the ten questions at the beginning of this chapter to identify your impact.
2. Choose your most important impact and be able to explain why it is important to you.
3. Identify at least one thing you could do in the next six months that would help you reach your most important impact goal.

Chapter 10
Step 2: Plan for impact

Summary

*In this step, you will become clear about who might benefit from
your research and how you can engage with them to achieve
impact, using an impact plan. Creating an impact plan using the
template in Part 4 of this book is incredibly easy. It may take
anywhere between an hour and half a day to come up with a first
draft, but investing time in this step will save days and days later by
ensuring you don't waste time on activities that don't actually help
you achieve your impacts. In taking this step, you will turn impact
from a possibility into an inevitability. Follow your plan and you will
achieve impacts.*

Most research funders now require researchers to articulate how
their research might generate impact as a prerequisite for funding.
Getting this right can be crucial to your chances of funding success.
Although it goes without saying that to be successful the research
that is proposed must be of the highest quality and novelty, and fit
with the aims of the funder. However, many funding panels are
faced with a limited pot of money and a choice between a number
of proposals that each have the same top-quality rating. In such
cases, the quality of your proposed pathway to impact can make the
difference between a project being funded or not. Many funding
panels include non-academic stakeholders, who may pay particular
attention to your knowledge exchange plans, and it may swing a
decision in your favour if your impact pathway is particularly
credible.

When funded, you need to revisit what you proposed and update
and expand it into a fully-fledged impact plan. This will help you to
organise, implement and track your knowledge exchange activities
and impacts throughout the research process. This will also help

with reporting to funders, and if you realise the impacts you set out to achieve, then this process will also help you assess the impact of your research in future years.

Pathways to impact: a best practice library

Funders provide guidance on writing these sections of proposals, but the advice tends to be quite generic, and there are few examples of good practice available. For many years, researchers have asked research funders and universities to make examples of good practice available to help in the process of writing a strong pathway to impact. As a result, some institutions now make them available to researchers within their own institution, but not all researchers have access to this sort of help.

When I proposed developing an open access 'good practice library' of pathways to impact, I received a lot of negative reaction. Some people expressed concern that researchers would be tempted to copy examples of good practice, stifling innovation and leading to a narrowing of the pathways proposed for future projects. Others pointed out the fact that no example would be faultless, and so researchers might replicate elements of poor practice in addition to the good parts. Others still felt that the impact sections of a proposal should never be viewed in isolation from the wider proposal they are linked to. And what if good practice evolves? There is certainly evidence of this already. The examples of good practice given out by the UK Research Councils when they first introduced pathways to impact in 2008–2009 are now regularly identified by researchers in the workshops I run as examples of poor practice, compared with more recent examples.

My response was to use this negative feedback to design a best practice library that dealt with each of these issues. You can find examples on the Fast Track Impact website from multiple disciplines and funders, regularly updated as our perceptions of good practice evolve, accompanied by the summary and objectives of the research they relate to, with feedback from grant reviewers on their impact (where available) and a narrative I have written identifying elements of good practice and areas where I think they could have done better, so you can learn from both the successes and limitations of the examples provided.

Contrary to the view that sharing good practice stifles innovation, I would argue that open access to good ideas enables the best innovations to spread widely and be improved upon rapidly. Pathways to impact are so specific to individual projects that anyone attempting to copy these sections of a proposal wholesale will inevitably reap negative consequences in review, based on the poor fit to the research, even if no one spots the plagiarism. There are certain knowledge exchange mechanisms that have now been clearly shown to deliver impact better than other mechanisms, such as the Stakeholder Advisory Panel (detailed in my guide in Part 4 of this book). If these mechanisms have been proven to work, why are we trying to keep them a secret from others? Are we more motivated by institutional competitive advantage or learning from each other to generate impacts that work?

My best practice library is still small, but you can help by proposing examples to me. I will evaluate your example and if I think it is good practice I will include it in the library:

- Access my best practice library:
  www.fasttrackimpact.com/pathways-to-impact
- Contact me to propose your own examples of good practice:
  www.fasttrackimpact.com/contact-us

What does a top pathway to impact look like?

You can read detailed guides to writing the impact-related sections of your next grant proposal tailored to the needs of specific funders (e.g. UK Research Councils and EU funders) at: www.fasttrackimpact.com/resources. Here, I want to summarise the key points that I think are important across all research funders.

I can boil down my advice on writing the impact sections of your grant to the following: **be goal-driven and specific.** Proposals that seek to achieve a clear impact goal are more likely to inspire reviewers and be believable, rather than falling into the trap of simply communicating research findings, or listing numerous activities that have no clear point. For me, specificity equals credibility. A proposal to work with 'policy-makers' requires a far greater leap of faith on the part of reviewers than a proposal to work with a named policy team within a specific government department. The fact that you know the names and positions of key people who are in a position to enact change implies that you have a credible plan that will work. Identify your goals clearly, with specific indicators that will tell you when each one has been met. Explain how you will complete each activity in credible detail and why this is the best way of achieving a particular impact, e.g. instead of listing social media, identify the platform you will use, who you will target on that platform, and what impact goals you will be able to achieve via this medium.

In a nutshell, then, here are the most important things you need to get right if you want to write a research proposal that will have impact.

1. Specify inspiring and yet credible impact goals: see the questions in the previous chapter for identifying impact goals, and make sure you are aiming to achieve tangible benefits that someone somewhere will value, rather than developing communication or interim goals that aren't associated with explicit benefits.

2. Specify beneficiaries and the benefits they will get from your work (and those who may be negatively impacted by your work): skip to Chapter 14 for methods you can use to systematically identify publics and stakeholders who are most likely to benefit from your research and/or influence your ability to achieve impacts. Bear in mind that an impact for one group in one place or time may be negative for another group at another place or time. Consider how certain groups might empower you to achieve more impacts more effectively and efficiently, and how other groups might oppose or block your impact, and consider how you might mitigate negative outcomes for these groups, so you can work together. Make sure that you are able to articulate exactly what benefits each group will get from your work, rather than just making a long list of interested parties.

3. Demonstrate demand for your work: find evidence to show that there is an urgent and important need for your work to meet non-academic needs and priorities. Co-produce your pathway to impact (and elements of the research) with beneficiaries and consider establishing an advisory panel (see my guide in Part 4 of the book) where they can continue to guide your work to ensure its relevance and impact. If your impact includes commercial exploitation, work with industrial partners to do the necessary market research prior to submitting your proposal, so you can propose a credible commercialisation strategy. If this is not possible, get letters of support from major players in the sector showing support for the research and interest in the market. Also, make sure you cover any issues around intellectual property and regulatory and other barriers to market.

4. Map activities to your impact goals: systematically check whether you have activities that will take you to each of your impact goals, and that you have identified activities that match the needs and preferences of each public/stakeholder group you identified in your impact summary. Where possible, focus on two-way engagement with publics and stakeholders rather than one-way communication of findings, so you get feedback and can adapt your approach to be as relevant and useful as possible.

5. Establish your impact track record: talk about your track record on achieving impact, ideally with the groups and issues linked to your proposal. It is difficult to 'prove' that you will be able to do what you are suggesting, and some of the best evidence you have is a track record of having delivered impacts for these groups in these areas in the past. If you haven't got a track record yourself, consider bringing someone into your team who does and get them to work with you on your pathway to impact.

6. Build in impact evaluation: have a plan for evaluating whether or not you are moving towards or away from impact, which will tell you when you have achieved your goals. The process of identifying indicators will help you determine clearer and more credible impact goals. Thinking in detail about how you will know if you achieved impact will often identify risks and challenges that you can prepare for, making your plan even more credible. You can build any costs of monitoring and evaluating impact into your proposal.

7. Cost it: cost your pathway to impact and justify your request for these resources. This shows how seriously you are taking impact, and adds credibility to your claim that these activities will actually happen. Consider funding professional research communicators and impact experts, ensuring the funds allocated to this are proportionate to the scale of your project. Also consider including further training for your impact team as part of your proposal.

8. Weave impact into your research plan: if possible, weave your pathway to impact into your research plan, cross-referencing to it from your research at relevant points. Consider how you can work with your research funders to amplify messages and co-ordinate impact activities across other work they have funded.

9. Keep it simple: use plain English and make your pathway to impact stand alone (e.g. by spelling out acronyms), as a lay member of a funding panel may only read the impact-related parts of your proposal in any detail.

10. Seek specialist impact pre-review feedback: don't rely on academic pre-reviewers to provide feedback on the impact sections of your proposal. Instead, seek feedback from someone in your university who specialises in impact or, if possible, get feedback on these sections from someone who works with the publics or stakeholders you want to benefit.

Common mistakes in the impact sections of grant proposals

I've reviewed proposals for five out of the seven UK Research Councils and sat on funding panels for EU and overseas national research funders. Here are a few of the most common mistakes I have seen in the impact sections of proposal:

- No clear impact goals (or the goals are just about communicating the research to stakeholders or publics)
- Benefits for researchers and the academy are included in the non-academic impact sections of the proposal. Most commonly these include training and career benefits for early career researchers and students, and conference and workshops that will mainly be attended by researchers. If you genuinely want to include capacity-building for your research team or students as part of your impact, explain how they will be able to use their skills and experience outside the academy to generate societal or economic benefits, and consider how you will achieve these benefits at scale, and evidence that they actually happen
- Social science data collection methods are replicated from the case for support in the pathway to impact, claiming that the knowledge or engagement gained from these methods will generate impact
- Public engagement for the sake of it — you have a clear pathway to impact via policy or industry and the reality is that your work is so niche that very few members of the public would be interested, but you're going to bore the socks off a bunch of unsuspecting passers-by because you felt you had to add public engagement into your pathway to impact
- Vague plans lacking detail, which are rarely credible

Finally, many people remove any impact goals and associated activities that are uncertain or high risk, leaving only a small number of highly conservative outcomes and activities, which fail to inspire or excite reviewers or panel members. Your funder will not expect to see every goal achieved in the same way as your research objectives, so the risks of dreaming big are relatively low, and the higher you aim, the higher you are likely to reach. You should, however, only ever promise to do things that are credible and feasible, and which you actually intend to pursue.

Develop your impact plan

Whether or not you need to submit your impact plan to a funder as part of a research proposal, it is important that you have a clear plan that can guide your own progress towards reaching your impact goals. I use an impact planning template to organise my thinking clearly. You can see the template I use in Table 3 (download an editable version and a worked example from my website at: www.fasttrackimpact.com/resources). As a 'logic model', the template is driven by the impact goals you established earlier in this step. If you are still not convinced by your goals, dive into the rest of the template and revisit your goals at the end. In my experience, the process of thinking through your pathway to impact in this structured, detailed way usually brings the clarity you need to reframe your impact goals and make them convincing and powerful.

I have written about the methodological roots of this template in a peer-reviewed article in the journal *Evidence & Policy*. It is adapted from a tool called 'Logical Framework Analysis', but after attempting to train academics in this approach I discovered that people got so hung up on differentiating between inputs, outputs, outcomes and impacts that they made no progress (partly perhaps because the definition of impacts versus outcomes in this approach is not theoretically sound). I wanted to develop an approach that was quick and easy for busy researchers to use, and that specifically focused on research impacts, rather than the many other impacts we could derive that may not be linked to our research. I have now taught this approach to thousands of researchers, and many more who have read my publications have used it.

Table 3: The Fast Track Impact Planning template (for a worked example and an editable version see www.fasttrackimpact.com/resources)

Impact goal	Target stakeholders or publics	Reasons for being interested in the project	Activities to engage this target group	Indicators of successful engagement [and means of measurement]	Indicators of progress towards impact [means of measurement]	Risks to activities [and mitigation]	Risks to impact [and mitigation]	Who is responsible and what resources are needed?	Timing

Once you've identified your impact goal(s) in the first column, the next task is to do a basic publics/stakeholder analysis:

- **Identify who is likely to be most interested** in your research, describing the different motivations, needs and interests of each group in as much detail as you can.
- **Consider which groups are likely to benefit most** as a result of engaging with your work, and in what ways.
- **Consider which groups are likely to have most influence** over your ability to complete the research and generate your intended impacts.

My template specifically encourages you to identify which parts of your research each group is interested in, so you ensure that the pathway you develop derives impacts from your research, rather than from other sources of evidence. You can, of course, include these wider impacts, but you should be aware that these impacts may not be research impacts (depending on the nature of the evidence they build on) or they may not be *your* research impacts. For more information about the importance and benefits of this task, see the second principle (Chapter 5), and for detailed methods and a publics/stakeholder analysis template, see Chapter 14.

Once you've identified and analysed your publics and stakeholders, you can **identify activities** that are appropriate to use with different groups, to communicate or (if possible) co-produce messages from your research. Tailor these activities to the needs and preferences of each group, recognising that there may be different sub-groups within any single public or stakeholder group which may want to engage in very different ways with you. Pay particular attention to activities you may need to develop for influential and/or hard-to-reach groups, as these may take more time and effort, so you might want to start engaging with these groups early.

Identify indicators or targets that you can use to track whether or not your activities are actually taking you closer to your impact goals. Use this information to adapt or change your activities, so that you ensure you achieve your goals. To make this more powerful, I like to identify activity indicators that specifically tell me how each of my activities is working, providing me with formative feedback on my practice. I then separately identify impact indicators that will tell me whether or not my intended impacts are happening as expected (e.g. milestones have been reached) or I have actually achieved the impact. Before doing these two things separately I used to find that I had often only identified activity indicators (because they were easier to develop) and I had no way of knowing whether or not I was actually achieving impacts. I have also asked you to provide a 'means of measurement' so that these indicators are concrete and feasible to evaluate (whether in quantitative or qualitative terms). Prior to this, I was finding that many of the indicators I developed, while highly accurate, were too costly or time-consuming to be put into practice. This attention to detail at this point in the impact plan can also really help bring impact goals into sharper focus, so consider revisiting your goals once you've developed indicators to see if you can further improve them. For more help on identifying indicators, see my guide to identifying impact indicators in Part 4 of this book.

Consider the risks associated with achieving your intended impacts, and the activities you have chosen to reach them. What might not work or go wrong? Might there be unintended consequences? How can you mitigate these risks? I have seen research proposals being sunk at funding panels because they did not seem to be aware of some major risks that appeared obvious to the panel. If the researchers had identified the risks and explained how they would mitigate them, they might have stood a chance of being funded. However, appearing to be unaware of the risks can undermine confidence in a team's experience and credibility. It is better to identify risks yourself and explain how you will mitigate them than it is to hope that the reviewers and panel members won't notice them.

Ask what resources or help might you need to achieve your impacts and mitigate these risks. Consider who will be responsible for each activity and when you will time these activities in relation to your research programme and the priorities and agendas of your stakeholder community. Some activities and impacts may be impossible to realise before certain research tasks have been completed. Sometimes it is possible to identify a key date before which preliminary findings could be put to particularly good use (e.g. as part of a policy consultation), and it may be worth considering whether your research schedule could be adapted to provide results in time to be useful for that purpose.

Decide who is doing what and when. Decide if there are any deadlines or target dates for any of your activities or impacts, and assign responsibility for activities to relevant team members. Talk to your colleagues and get their feedback on your plan for impact, and if you can, get feedback from stakeholders so that they can tell you if there are particular points at which certain findings could have greater impact. Then make a habit of checking in with these people. Regularly update yourselves as a research team on your progress towards impact goals as well as updating yourselves on the progress of the research itself. Put it on the agenda of any regular meetings you have with colleagues. And consider how you can create an accountable and collaborative ongoing relationship with the stakeholders who help you make this plan, for example, through a bi-annual stakeholder advisory meeting.

Finally, it is worth noting that there is a danger that impact plans can become too prescriptive. Targets and indicators can help keep your impacts on track, but they shouldn't become a straitjacket that prevents you from adapting your objectives to meet changing stakeholder needs or exploiting new opportunities as they arise.

Your tasks for this step

1. Turn your impact goal from the last chapter into an actionable impact plan. The first step is to identify publics and stakeholders who are likely to be interested in your work, to benefit from engaging with you and/or who can influence your ability to generate impact.
2. Identify activities you can use with different publics and stakeholders and tailor your activities to their needs and preferences.
3. Identify indicators or targets that you can use to track whether or not your activities are actually taking you closer to your impact goals.
4. Consider the risks associated with achieving your intended impacts, and the activities you have chosen to reach them. What might not work or go wrong? Might there be unintended consequences? How can you mitigate these risks?
5. Ask yourself what resources or help you might need to achieve your impacts and mitigate these risks.
6. Consider who will be responsible for each activity and when you will time these activities in relation to your research programme and the priorities and agendas of your stakeholders and publics.

Chapter 11
Step 3: Cut back anything hindering or distracting you from your impact

Summary

In this step, I want you to think about how you can make room in your busy schedule for impact-generating activities that are based on trusting relationships with the people who might be interested in and use your research. Relationships take time, and time is in short supply for most researchers, so getting this step right is essential. If you complete this step, then it won't just free up time for impact, it will transform your work-life balance too.

Time is something that is in short supply for most researchers. That's a problem if you want to take a relational approach to impact, because investing in relationships takes time. That's why this step is designed to help you become more efficient and focused as a researcher, so you can make time for relationships with those who are interested in your research (and those other important people in your life, whether or not they are interested in your research!). I have written more extensively about this in *The Productive Researcher*. Here, I want to summarise some of the key lessons from that book and point you to some additional ideas that can significantly change how you work.

Prioritise

How to prioritise is the single most important lesson I have learned from my reading, interviewing some of the world's most productive researchers and my own experience. You might think you already know how to prioritise, but the key thing that seems to set the most productive researchers apart from the rest of us is their unusual ability to retain a laser-sharp focus on their priorities from minute-to-minute, day after day and month after month. This unusual level of focus comes from the way these researchers set their priorities. These priorities are more than simply preferences; they express the values and identity of the researcher. Priorities that come from this deep place stand the test of time and are deeply felt. As a result, they motivate researchers to retain focus on their priorities, making it easier to say "yes" to the things that are most important to them, and "no" to many of the things that might otherwise have distracted them.

Be mindful

Start trying to be mindful of how you spend your time every day, and to what extent each of the things you are doing connect with your impact goals, and your wider life goals. In this step, you are going to identify at least one regular task or activity that you can either entirely cut out of your schedule or drastically cut back on.

Think about how you work with everyone on your team and whether there are people who might benefit from some of the more minor tasks on your to-do list, for example, getting experience reviewing papers that can help them learn how to write more effectively themselves. Consider getting a virtual personal assistant (PA). I have a UK-based virtual PA who has experience working with academics, though there are more generalist virtual PAs who charge less. What would you say if I gave you the choice of a pay rise (say £300 per month) or a whole extra day per week to do whatever you want (you can use it to achieve more in your work or spread your work out over the week so you don't have to work so hard — it's up to you). For most academics, a whole extra day per week is worth much more financially than £300, let alone psychologically. My PA helps me format reference lists, does simple research tasks for me (e.g. finding out how to make a MOOC), writes up notes (including Post-it notes and flip-chart paper) from workshops, sets up meetings, organises all my travel and expenses — the list is endless. So next time you get a pay rise, have a think about how you can turn that small amount of extra money into a significant amount of extra time.

Lack of time = lack of priorities

I personally believe that the problem all academics will recognise, of the day never being long enough to do everything we are meant to do, comes down to poor priorities. Lack of time = lack of priorities.

Perhaps you are regularly working evenings and weekends, and don't have time to think (let alone do anything) about generating impacts from your research? I've discovered something interesting over the years, which I've recently found there are names for: Parkinson's Law and the Pareto Principle.

Parkinson's Law simply recognises that tasks will swell to fill the time you give them. Therefore, you need to limit the time you give to the tasks you need to do. The strange thing is that the end product is usually as good as if you'd spent double the time on the task, and sometimes actually far better. The reason for this is that the level of focused attention provided by a forced deadline actually enables you to produce more focused work.

I first discovered this in my teaching when I was told I had to head up a research centre, and had significantly less time to prepare my lectures. Without intending to do this, I ended up regularly preparing my lectures on the day of delivery, and although it was more stressful, I was surprised to discover that my ratings from students increased significantly. Admittedly, I had a fairly good grasp of the subjects I was teaching, but the key difference was that I was now having to rely less on my notes and more on my intuition, and as a result my passion for the subject became far more apparent, and this enthusiasm was infectious. I also did more class exercises rather than boring people with information overload, which meant that the students actually learned more. I did the same with a literature review, giving myself just one week to do all the reading and the writing of the first draft. It was a hard week's work and I made lots of edits and changes before the final version was eventually published. But this is now by far my most cited paper (>2000 citations) and has played an important role in establishing my reputation in that particular field.

What 20% of your working day produces 80% of the outputs you value most?

The Pareto Principle suggests that for most people, 80% of the outputs you value most come from only 20% of the time you spend working. In reality, it is not exactly 80:20, but I believe that the principle holds for most of us. I think that this principle is the source of many researchers' greatest frustrations and disappointments, because we spend so long every day doing the urgent things that everyone around us is shouting for at the expense of the less urgent, but far more important, things. So that paper or book we dreamed of writing stays unwritten, and we accumulate regret and frustration for the sake of keeping everyone around us happy. I'm not saying that we should stop being team players and be selfish with our time. But if you've got a really important goal in your work, try and spend some time on it every day, even if you only manage half an hour, and you'll be amazed by how much more satisfied you feel day after day, week after week. Some people I know get up at five or six in the morning to spend an hour writing or whatever that important task is. I've never been motivated enough to do that, but the point I'm trying to make is that you don't have to spend all day doing the important stuff, but you do have to keep chipping away at it. When you start focusing on the important things, you suddenly realise that many of the urgent things you're being told to do aren't actually that important. With this revelation it becomes easier to say 'no' or take a few shortcuts to do a 'good enough' job on those tasks, so you can get back to the really important stuff.

So what 20% of your working day produces 80% of the outcomes you value most? Write a list of everything you did yesterday and, if possible, estimate how long you spent on it. Be brutally honest about how long you spent replying to emails, on social media and doing other things that yielded very little tangible outcome. For me, this isn't necessarily about cutting these tasks out — you're not going to be popular with your students or colleagues if you stop replying to emails. But it is well worth considering how you could be more strategic.

What could you cut?

Here's a list of the things I have drastically reduced time on, which might inspire you to consider what you could cut:

1. **Social media:** I no longer try and read everything in my feeds, and limit myself to a 30 minute 'news' window every day which I spend reading news via Twitter. In addition to mass media outlets, I'm following a targeted social media strategy designed specifically to help me achieve impacts through my research (see Chapter 17). Now I'm not blogging into thin air — I've got a strategy to drive traffic to my most important research outputs and to engage with specific audiences around key messages that link to the impacts I want to achieve.

2. **Email:** I do a scan of the most important and urgent emails in the morning and only reply to these, ignoring the rest till the afternoon, often after lunch when I'm feeling most tired. If necessary, I will send a holding email, acknowledging the message and explaining that I've got a busy morning and that I'll reply later in the day. If I've got a writing deadline (even if it's self-imposed), I will put an out-of-office reply up, explaining that I'm working towards an important deadline and not checking emails for the rest of the week, and that they can send me a text message if it is urgent, otherwise I'll get back to them as soon as I can.

3. **Reviewing:** none of us can escape this; it is part of our duty to review others' work in as timely and constructive a way as we would hope others would review ours. However, I think we sometimes feel a false sense of duty to review more than we need to. I personally feel that I have met my moral obligation to the academic community if I review around three times as many manuscripts and grant proposals as I submit myself (on the basis that most of the papers and grants I submit will get three reviews each). In addition to only reviewing papers that are in my subject area, I will only review papers from which I think I'm likely to learn something useful or new (based on the abstract). This rule also has a handy way of ensuring you don't end up reviewing papers for 'predatory' journals.

4. **Committee meetings:** there is usually someone absent at each committee meeting you go to, and no one really minds, as long as they aren't always absent. What would happen if you decided to make your apologies for every other committee meeting or shared the role with a colleague so you could take turns? What would actually happen? Would the meeting cease to function? If you missed out on a key piece of information or decision, would there be no other way to find this out or influence that decision? Obviously, it is wise to check the agenda and if you're chairing the committee then this won't be possible, but if you're not chairing it, do you really have to be at every single one?

5. **Chat:** it is incredible how much time is wasted just chatting in the corridors, or with people randomly passing your office door who want to say 'hello'. Rather than leaving your office door open with an invitation to be interrupted, make yourself fully available to anyone by appointment, and then stack all your appointments into a single day or a couple of days in the week. This doesn't mean that you need to become a hermit and avoid all social contact, but now you can be strategic and targeted in who you actually invest time in, whether professionally or as friends, by taking these people for coffee or lunch, where you can have quality time together. Since my PhD, I have always worked primarily from home, coming into the department on selected days for meetings and to socialise. The rest of the week is for being productive and focused.

Those are just a few of the things I've done over the years to increase my productivity. I'm not suggesting you should cut back in the same areas, but hopefully this list gives you a flavour of the sorts of things that might be possible.

Do less to do more

Many of us think that somehow the world will end if we stop doing some of the things that all other academics do. But ask yourself the question: what would actually happen if I didn't do this? Really? Think of your worst-case scenario, and then consider if you could reverse, cope with or recover from that situation. If you'd manage to cope with the worst-case scenario, then go ahead and cut it. In almost every case, that worst-case will never actually come to pass.

We all know that there will always be many more things that we 'ought' to be doing every day, which we run out of time for. The art of being a successful academic with some semblance of work-life balance is more about what you choose not to do, than what you do. The more time you give to your work, the more tasks will fill up that time. That's why after my PhD I made the decision to never work weekends, and only to work evenings when my wife or I were travelling (typically two evenings per week). I feel rested and refreshed on a Monday morning and I don't resent my work. That simple act of constraining my time is, I believe, one of the reasons I have been able to be so productive.

My message is this: do less to do more. Limit your tasks to only the most important, so that you shorten the amount of time you have to work. Then, shorten the amount of time you have to work, so that you are forced to limit your tasks to only the most important.

Tasks for this step

1. Make a list of everything you did during your workday yesterday (or your last typical day in the office) and how long you spent on each task, including menial and non-work tasks.
2. Commit today to removing one thing from your schedule to make room for impact-generating activities. If you can, commit to removing as many other things as possible, so that in addition to being more productive, you can get a better work-life balance (which, of course, will make you more productive in the hours you do work).

Chapter 12

Step 4: Get specific about your impacts and the people who can help you

---

Summary

Your task in this step is to come up with one thing you can do this month that will take you closer to achieving impact. In parallel with this, and to help you actually do this one thing, you will identify key individuals who can help you achieve the impacts you want to see.

Now you have cut back anything that might be hindering or distracting you from your impact (Step 3, Chapter 11), you can revisit your impact plan (Step 2, Chapter 10) and start to get specific about the impact and activities you want to prioritise for action this month.

Whose footsteps are you following in?

Being successful at something is much easier when you are surrounded by other people who are already being successful at that task. Being connected to other people who are on the same journey towards impact as you, who already have some experience under their belt, means you can call on each other for help and advice. It is also motivating to see what others have done, and to see how they achieved those successes.

These people will probably be other academics, but they may be knowledge brokers from charities, industry or elsewhere. They may already be achieving some of the sorts of impacts that you would like to be able to achieve yourself.

What they will have in common is that they are a few steps ahead of you on their pathways to impact. They may be ahead of you in terms of their experience generating impacts generally, or specifically generating a particular type of impact or using a particular technique, technology or activity that you would like to be able to use yourself.

It is surprising how open many people are to being contacted by someone who wants to learn from them. If this is done in a spirit of humility, most people will respond positively and be willing to mentor you in some shape or form. Even as a PhD student, I reached out to key authors in my field to co-author papers as part of my PhD, and got incredible mentoring from these researchers in addition to the inputs of my PhD supervisor (and ended up publishing 12 papers out of my PhD, over half of which have been cited over 100 times). So don't be shy!

Create your influence network

In a moment, I'm going to ask you to identify five people who are a few steps ahead of you on their pathway to impact who might be able to mentor you. Next, I'd like you to identify five more people. This time, rather than looking for mentors, I'd like you to look for influential individuals who are not researchers and who might have the power to enable you to have significant impact. They may have knowledge and contacts that could significantly help you to achieve your goals. They might have resources at their disposal that they want to use to achieve similar goals to yours. They might have significant followings on social media that could enable you to get your message across to a wide audience. They might have access to data, hard-to-reach groups or sites that you need to complete your research. You might already have identified some of these people in your publics/stakeholder analysis in Step 2 (it is worth going back and looking at the groups you identified as being particularly influential). These people are worth their weight in gold, so it is a good idea to spend time talking with colleagues and trying your best to identify these individuals, so you can begin to connect with and draw upon their influence to help you achieve your impacts.

Tasks for this step

1. Identify one impact you could work on this month. This could be your most important impact, identified in Step 2, the goal you identified in Step 1 that you could do in the next six months, or it could be another impact, but ideally it should link in some way to your most important impact. Then identify a specific activity that will help you achieve that impact (even if it is only the first of many steps you will have to take to reach the impact).

2. Make a list of five people who you feel are ahead of you in achieving impact or working with the end users of research — these people will probably be other academics, but they may be knowledge brokers from the third sector, industry or elsewhere.

3. Make a list of five stakeholders from Step 2 who are likely to be particularly influential in enabling you to complete your research successfully and make an impact.

4. Make a commitment to reach out to at least one person from each list this month to try and establish a working relationship that can help you achieve impacts from your research. Prioritise those who might be able to help with the one thing you've decided to pursue.

Chapter 13

Step 5: Achieve your first step towards impact and monitor your success

---

Summary

Finally, in this chapter you will take your first step towards achieving impact and start monitoring your success. This isn't about achieving your impact now (though in some cases this may be possible), but about taking purposeful, measurable steps each day that take you closer to specific impacts you are targeting. Focus on small steps every day and celebrate your progress.

If you have completed the tasks at the end of each of the preceding steps, then you have already invested wisely in yourself and are a huge step closer to being able to generate significant and far-reaching impacts from your research. You now have a much clearer idea of the impacts you want to achieve and why. You've got a clear plan that can get you from where you are now to the impacts you want to see in the future. You've cut back on the things that have been holding you back to make more time for yourself and more time to generate impact. And you've identified a network of people who can help you reach your goals

During the last step, you identified one thing you could work on this month, which could take you closer to achieving impact. During this step, you will actually start work on this. It is important that you actually do something tangible as part of this step. You've done a lot of thinking, discussing and planning so far, but taken little real action. Now is the time to put everything you have learned into practice.

Put theory into practice

In the last step, you identified one thing you could work on this month. If it isn't particularly tangible, then your first task is to find a way to make it more so. Is there a way you could involve others to do this with you, for example, in a workshop setting with stakeholders? Is there some sort of physical artefact that you could produce, linked to the thing you've decided to work on, like a policy brief, a film or an educational resource?

Next, go back to your impact plan in Step 2, and think as deeply as you can about the activities you will undertake, how they link to your impact, how you can adapt your activities to the needs of different publics and stakeholders, and the risks that might be associated with those activities. If you haven't already done so, get some feedback from colleagues about what you are planning to do, to see if they can spot any flaws or limitations in your plan.

Next, go to the people you reached out to in Step 4 — they should already be waiting for your email or call if you set this up right when you contacted them. Although it might feel uncomfortable at first, your task is to ask them for help. Ask for something very specific that you think they should in theory be able to give you. If you don't ask, you will never know if they would have been happy to help you, and more often than not, people actually want to help.

Finally, just go out and do it — whatever the thing is that you decided to do, and make sure you do it within four weeks so that you can maintain momentum.

Double your failure rate

There is one last thing you need to consider though. It probably feels great that you are actually out there doing something tangible to generate impact. But how do you know that what you have done has actually effected change, and helped you get closer to your impact? In the impact plan I mentioned in Step 2 (see the template in Part 4) there's a column where you have to identify indicators or targets. Make sure that you design whatever it is that you do this month so that there is some way of collecting information about whether or not it is working. If it didn't work as well as you wanted, do something else next month. Those who succeed most in life are often those who are prepared to experience failure again and again, and it is for this very reason that they learn how to succeed. As the American businessman and philanthropist Thomas Watson Jnr put it: *"if you want to increase your success rate, double your failure rate"*.

For many of us, this requires a very different perception of failures, not as things to be avoided at all costs, but rather as inevitable bends in the road to success, from which we can learn important lessons. Learning to accept and embrace failure takes a degree of self-confidence that I only achieve on good days. I think that one of the reasons that failure is so hard to accept as a researcher is the level of personal creativity we invest in our work, and the fact that we are our own brand with a reputation that is inextricably linked to our name. For this reason, it is difficult not to take professional rejection personally. However, this can be the beginning of a downward spiral. The most common cause of writers block among colleagues I've been asked to mentor is lack of confidence. And my own periods without research funding have been prolonged by doubt that anything I could ever submit would be fundable. Many of us are tormented by imagined criticisms of reviewers before we even complete the first sentence of our paper or proposal.

The trick, I believe, is not to ignore those voices but to embrace them, and tackle each criticism constructively as it comes to mind. When you call these criticisms to mind consciously and make them explicit, you can start to distinguish between genuine insights about weaknesses in your work (which you can address) and the inner psycho-babble that (in my case) tells you you're worthless and can't do it (which you can also address, but in different ways). Rather than hearing these reviews as criticism and allowing them to cripple you with low self-esteem, work hard on turning these internal voices into something that can propel you forward. I think it is hardest when you know you messed up. With hindsight, you can see what mistakes you made that led to failure. However, it is still possible to learn from these mistakes, if you don't take your failures personally and pick yourself up and try again. Andrew Derrington, in *The Research Funding Toolkit*, tries to help by conceiving of research as a 'grants factory', in which researchers churn out proposals dispassionately on a production line, starting work on the next proposal as soon as the last one is submitted, and accepting the odds that if your work is any good then eventually one will get funded. Whether or not you are able to detach yourself from your work to that extent (I'm not sure I can), I think that there is something to be said for just picking yourself up and carrying on, no matter how bad your failure.

I've experienced a number of talks in which I 'went down in flames' — usually as a result of my own lack of preparation rather than any killer question. I think the most acutely embarrassing was a particularly high-profile conference that I'd been told I had to present at by my funders. I knew that I had to find and impress the head of the Commission for Rural Communities as he was commissioning work for an inquiry that would influence policy. The great and the good, including the head of the organisation that had funded my research, were sitting in the front row, and I was nervous. I finished my talk about UK uplands, and the head of the Commission stood up to ask the first question. As I tried to make a mental note of what he looked like, so I could find him later, the pressure suddenly became too much for me. He had asked the most ridiculously easy question, and it was the simplicity of the question that put so much pressure on me; I knew that a strong answer to his question would easily convince him that we should do the work for his inquiry. But my mind went completely blank. I could think of nothing — absolutely nothing — to say. Sir Howard Newby, then head of the Higher Education Funding Council for England, was chairing the session, and he politely looked at me and said into his microphone, "I think this one is for you Mark". All I could say was, "I know". After what felt like an age, Sir Howard took another question and someone else on the panel of speakers answered, while I wished I could magically disappear. To make matters worse, I thought I could redeem myself by asking an intelligent-sounding question of the speaker who followed me. I finished my question, feeling like I might have regained a tiny amount of credibility with the audience, and Sir Howard intervened, saying "You do realise that the legislation you're asking about applies only to uplands, and this research is all about lowlands, don't you?". I could have died of embarrassment as my fellow speaker came to my rescue with an answer to my unanswerable question.

We've all been there. Everyone has got stories like this, which we don't typically tell (let alone write about in books), and because we don't tell our failure stories, we end up with the perception that we are the only people who fail, and that everyone else moves from one success story to the next. I think we need to be brave enough to start laughing about our failures so we can all learn from each other's mistakes and move forward together more empathetically to achieve impact, no matter how many failures it takes along the road.

Focus on the small steps that take you closer to impact every day

Make it your goal to take small steps every day that you believe will contribute to your impact. At the heart of this book is a relational philosophy of impact, which focuses on building empathic, trusting and respectful relationships.

Rather than focusing on the end point and beating yourself up for your slow progress towards your goal, focus on repeating small, value-orientated behaviours linked to your goals, and trust that these will take you where you need to go. This doesn't mean you don't monitor and correct your course — but rather than focusing on whether you've reached the top of the mountain yet, celebrate that you took a few more steps that were in the right direction that day.

Finally, if you have evidence that your activities are working, or even better, that you are actually achieving impacts, then shout about it!

Celebrate and share your success with others who are on a similar journey to you, so they can be inspired and learn from you.

I showed how you can integrate monitoring into your impact plan in Chapter 10, and I'll give you more detailed information on how to track, evaluate and evidence impacts in Chapter 22.

Remember, if you want extra help with all five steps you can sign up to receive each step via email every week over five weeks on the Fast Track Impact website at: www.fasttrackimpact.com/for-researchers. I'll talk you through each step in a short video.

Tasks for this step

1. Start working on the impact you identified in the previous step — make it tangible and actually do something. Now is not the time for planning or talking about it. Now is the time to actually put your impact plan into action.
2. Reach out to the people you connected with in the previous step and ask them to help you achieve this impact with you this month.
3. Make sure you collect information that can tell you whether or not your activities are actually taking you closer to impact.

Part 3: Tools and techniques

Chapter 14

Prioritising stakeholders and publics for engagement

I explained the importance of systematically representing the needs and interests of relevant publics and stakeholders in the 'represent' principle (Chapter 5). To enact this, the second step to implement these principles involves doing a publics/stakeholder analysis. This chapter explains how to do this.

For me, the power of a publics/stakeholder analysis is its ability to enable you to prioritise who you engage with first. If you have time, you can use this tool to derive long, comprehensive lists of all the publics and/or stakeholders who should in theory be interested in or benefit from your research. In my experience, such lists just intimidate me, as I know I will never have time to reach out to everyone that has been identified. Instead, use this tool to empower you to take strategic first steps towards engaging with the publics and stakeholders that are most important to you. Whether you have time to reach out to three or thirty contacts, you know that you contacted the most important groups when you run out of time to contact anyone else.

The two most common ways I prioritise using this analysis is to focus on the hard-to-reach groups first, or to reach out to knowledge brokers who can short-cut me to relationships with key people from across multiple social and professional networks (see the 'engage' principle in Chapter 6). You may want to prioritise those with greatest power to facilitate or block your impact, or you may prefer to prioritise marginalised groups. You make the choice based on your own preferences, and use this to take a strategic approach to who you engage with first. If you have limited time and are being approached by many different organisations, you can use this approach to justify putting off engagement with certain groups, or sending them to your social media feeds or newsletter, so you have time to reach out to your priority groups, and don't get side-tracked by constantly reacting to those who shout loudest.

Prioritising publics

The first thing I need to say about prioritising publics is that there is no such thing as 'the general public'. If you held a public engagement event in an open space near your place of work on a weekday, and surveyed those who came, you would find that certain people came while others did not. For example, you may find that you attracted people of working age, few children, more students and university staff if you work at a university, people within certain income brackets, who are more or less educated, with certain sorts of interests and so on. If you ran the same event at the weekend in a village, and surveyed the people who attended, you would find that you had attracted a very different group of publics, for example, more young families with different levels of income and educational attainment and quite different interests. Even television programmes reach specific publics based on their interests, education level and time of life. Rather than engaging with the general public, you will always engage with specific publics.

Your challenge is to establish which publics are likely to be most interested in your work, and which will benefit most from engaging with it.

To do this, you can use an interest-benefit matrix (Figure 5):

- In the top-right quadrant are your **easy-to-reach target publics**, who will benefit significantly from engaging with your research. They are particularly interested in your work, partly because they will benefit directly in some way. For example, the benefit may simply be an opportunity for them to indulge a personal interest and learn new things about your area of research. Alternatively, the benefit may be learning from your research about new ways of avoiding a threat that people in their position are exposed to, such as cyber-bullying. Reach out systematically in priority order — contact the first one on your list now.
- More problematic are the **hard-to-reach publics** who are disinterested in your research but could in theory benefit significantly from engagement, if only you could reach them. For example, your research on youth gun and knife crime may provide particular benefits to young people from certain backgrounds, but they may not be interested in what you have

to say on the topic. Your challenge is find out what would motivate them to engage.

- Less problematic, but worth being aware of, are the **easy-to-reach non-target publics** who may engage more than hard-to-reach publics but benefit less. This is only a problem when it lulls you into a false sense of security that your engagement is working, when in fact you are not reaching those who most need to engage with your work. For example, I did some work with a university that wanted to build relationships with the deprived communities that lived around the campus. It ran a series of events on campus, to show people that they were open to engagement, and wanted to share their space with their local communities. The events were very well attended, but when they looked at their data, they discovered that the majority of attendees were university staff and students who were very interested in the research, and not their target group. Work with non-target groups if you have time, but be careful not to focus on these groups at the expense of those who have greater need.
- Finally, there are the **other publics** that are not particularly interested in your work and are unlikely to benefit significantly from engaging with you. This technique is all about enabling you to prioritise your efforts so you can save time. These are the groups you can leave till last, that it won't matter if you don't ever get round to reaching.

If you want, you can take this a step further and group interested publics according to shared interests. There may be different publics in the top-right and bottom-right quadrants who are interested in different parts of your work for very different reasons. This then empowers you to develop messages and activities that are tailored to each of those different interests, so you increase the likelihood that you are able to provide them with the benefits they are looking for. For example, a researcher might develop messages tailored to the interests of different audiences to 'nudge' them towards changing attitudes and behaviours. In well-funded projects, this would normally start with work that is designed to understand the attitudes and behaviours of different groups in relation to the research area, for example, family planning or sustainable living. Then different messages are developed that are likely to be acceptable and attractive to those different groups.

The problem with this is that most research can only be communicated in a fairly limited number of ways, and there are only so many ways you can adapt messages from your research whilst remaining true to your findings. For example, messages around research on new family planning techniques could not be adapted to suit the attitudes of groups who oppose family planning on religious or moral grounds. On the other hand, during my research on peatlands, I've managed to adapt messages from this work to emphasise benefits for conservation, the economy and farming communities, depending on the interests of the different groups I've been working with. For me, there is nothing disingenuous about this. I am just emphasising certain aspects of the research. I am not hiding the other aspects or manipulating the findings to suit the audience. However, not everyone feels comfortable with this approach.

	Low ← Level of Interest → High
High	**Hard-to-reach publics** who are disinterested but could benefit significantly from engagement *Find out what would motivate them to engage*	**Easy-to-reach target publics** who benefit significantly from engagement *Reach out systematically in priority order – contact the first one on your list now*
Benefit	**Other publics** that have little interest and are unlikely to benefit much if they were to engage *Keep a watching brief as their needs and interests may change over time*	**Easy-to-reach non-target publics** may engage more than hard-to-reach publics but benefit less *Be careful not to focus on these groups at the expense of those who who have greater need*

Figure 5: Interest-benefit matrix showing how publics differ according to their relative interest in your research, and the likelihood that they will derive benefits from engaging with your work

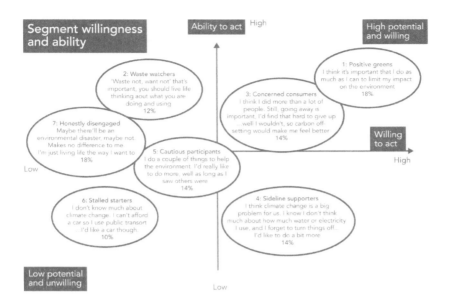

Figure 6: Defra's segmentation model divides the public into seven clusters, each sharing a distinct set of attitudes and beliefs towards the environment, environmental issues and behaviours, based on responses to a broad range of attitudinal questions as part of the 2007 Defra attitudes and behaviours survey (source: Defra (2008). A framework for pro-environmental behaviours. HMSO, London).

To illustrate how this can look, Figure 6 shows how the UK government's Department for Environment, Food and Rural Affairs segmented the public in relation to attitudes towards climate change to help them develop messages based on the latest research evidence that would be most likely to alter each group's behaviour towards adopting more pro-environmental behaviours.

Prioritising stakeholders

When I'm working with stakeholder groups and organisations, I make a small but important tweak to this approach. In addition to wanting to know which stakeholders might benefit most, I also want to know if there are any particularly powerful stakeholders I should know about. There are sometimes organisations that have the power to facilitate our impact. Their common interests and goals mean that our impact is their impact. As a result, they will throw money, staff time and data at our work to power us to impact. On the other hand, there may be organisations with the power to block our impact. For example, our research may threaten their profits, livelihood or ideology, and as a result these organisations may oppose our research as well as our attempts to generate impact. Identifying those with the power to facilitate or block our impact early is important if we want to harness the power of those who want to help us and understand the concerns of those who feel threatened by our work.

Figure 7: Interest-influence matrix used to identify stakeholders with differing levels of interest in and influence over your research

Now, instead of (or in addition to) looking at the relative benefit of my research to an organisation, I consider the relative influence (for good or for ill) of that organisation on my capacity to achieve impact. To do this, I use an interest-influence matrix (Figure 7):

- In the top-right quadrant, are the **easy-to-reach influential stakeholders** who could block or facilitate impact and typically engage easily and regularly with your work. You will probably know who these organisations are already (especially if they really don't like your work). It is not difficult to identify or engage with these groups because they are so interested in your work.

Reach out systematically in priority order — contact the first one on your list now.

- In the top-left quadrant are the **hard-to-reach influential stakeholders** who could block or facilitate impact but are not interested enough to prioritise engagement with you. For example, your work may be marginal to their interest, or be perceived as a narrow and hence minor angle on a bigger issue. This group is problematic because they may have the power to facilitate far greater impact than you could otherwise achieve, but may not be willing to use that power. Alternatively, they may turn out to be a gatekeeper. They are not interested enough to answer your emails or pick up the phone, and so you give up on trying to engage with them. However, later in your research, you discover that you can't do your work without them, and at that point you discover that they are blocking you from accessing the data, sites or people you need to work with. By that time, it may be too late to adapt your work to address their concerns.

- Then there are the **easy-to-reach but marginalised stakeholders**, who may want to block or facilitate impact but have limited influence or voice. These are often marginal stakeholders, who might warrant special attention to secure their engagement and to empower them to engage as equals in your research. However, the low level of influence held by this group is often used as a justification for excluding them from the research process. I like to identify organisations in this category that have similar interests to some of the more powerful organisations and engage them in the research through a stakeholder advisory panel, to try and form alliances with more powerful organisations. This turns the original purpose for which stakeholder analysis was invented on its head. Now rather than 'neutralising' stakeholder threats to company profits (as it was originally conceived to do), researchers are using these techniques to empower previously marginalised stakeholders to have a say in decisions and research that affects them.

- Finally, there are the **other stakeholders** with limited interest or influence. However, it is worth noting that their interest or influence may change over time, so it is worth keeping a watching brief.

The publics/stakeholder analysis tool

Although interest-benefit and interest-influence matrices are a useful way of learning about publics/stakeholder analysis, they have limitations. The most important issue I have is that a group of researchers may spend an hour discussing where different organisations might be placed in the matrix, moving organisations around as they discuss them. However, the only thing that is recorded at the end of this process is their eventual position. The rich discussion explaining what they were interested in or why they would benefit so much or have so little influence is lost. That is a problem because such information is extremely valuable for informing how you will engage with these groups.

For this reason, my publics/stakeholder analysis tool is an extendable matrix (Table 4) that asks you to document why you think each of your publics or stakeholder has a given level of interest, benefit or influence. It is extendable because it can be extended to consider a range of other factors that may help categorise and engage effectively with stakeholders, for example, identifying any important relationships between stakeholders (e.g. coalitions or conflicts), information about how best to approach and engage with different stakeholders, and contact information that can be used to check and further extend the analysis.

If you have time and resources, and are interested in knowing more about the social context you are going to be working in, there are a range of other methods that have been developed to understand relationships between stakeholders. These include ways to analyse the structure of social networks, to map stakeholder perceptions and values, and to assess and analyse conflicts between stakeholders (e.g. Table 4). Although these relationships may be used to categorise and prioritise stakeholders for engagement, these sorts of analyses are typically conducted after stakeholders have been categorised, to understand how different stakeholder groups interact with one another, and to identify specific individuals or organisations that may play an important role in diffusing knowledge or practices within and between different groups of stakeholders. Such methods can be useful to identify opportunities and risks of engaging with certain stakeholders, and identify the values and priorities of different groups, so that these can be taken into account in the design of an impact plan.

Finally, it should be noted that all methods for identifying publics and stakeholders provide a snapshot in time, and stakeholders and their interests and influence are typically dynamic. For example, organisations may form alliances to either promote or defeat a particular outcome and your publics/stakeholder analysis can be used to identify where such alliances are likely to arise. This requires your analysis to be revisited and updated periodically to ensure that the needs and priorities of all publics and stakeholders continue to be captured.

Table 4: Publics/stakeholder analysis tool

Name of organisation, group or segment of the public	Likely interest in your research. H/M/L	What aspects of your research are they likely to be interested in? Identify key messages linked directly to your research for this group	What level of influence might they have on your capacity to generate impact and/or what level of benefit might they derive from the research? H/M/L	Comments on level of influence and/or likely benefit (e.g. times or contexts in which they have more/less influence over the outcomes of your research, ways they might block or facilitate your research or impact, types of benefit they might derive from the research)	If influence or benefit is high but interest is low, how might you motivate greater interest and engagement with the research?

Figure 8: Images from a publics/stakeholder analysis conducted for the UK government's Department for Environment, Food and Rural Affairs, February 2016

How to do a publics/stakeholder analysis

I would recommend that you invite a small number of non-academics who know the stakeholder landscape well to help you with this task. But if you are short of time, then even if you just fill out the template in Part 4 with your research team, you will be able to do far more impactful research than you would otherwise have done.

The following methodology will take you approximately two days to complete, including between half a day and a day for an initial workshop (see the example facilitation plan in Part 4, illustrated in Figure 8), followed by a series of half-hour telephone interviews to check your findings with key stakeholders (which is also a great opportunity to get their feedback on the focus of your research and start getting ownership as you adapt your work to stakeholder interests). The following steps are designed to be straightforward and replicable, but this does not mean that they should be inflexibly applied. Local circumstances may require these steps to be adapted, to ensure that the publics/stakeholder analysis is a tool that brings stakeholders together and facilitates active engagement in research.

1. **Identify cross-cutting stakeholders:** Identify between two and four individuals from cross-cutting stakeholder organisations who operate at the scale of your research (if you have multiple study sites, you may need to do this for each site). The key criterion for selection is their breadth of interest in the issues you are researching, so that they are familiar with the widest possible range of organisations that might have a stake in your

work. Aim to represent a range of different perspectives on the issue so that you can facilitate debate about the relative interest and influence of different stakeholders (e.g. someone from a government department or agency and someone from an NGO, not just people from different government departments).

2. **Invite cross-cutting stakeholders to a half-day workshop:** only two to four stakeholders plus project team should be present as it is not the aim to represent all stakeholders at this workshop (this isn't possible as we have yet to systematically identify them). This workshop should take approximately four hours (half a day), but if there is time, it is more relaxed done over a day. During the workshop, complete the following tasks:

    a) **Clearly establish the focus of the research that you think individuals, organisations or groups might have a stake in:** it is important to be as specific as possible about your focus so you can clearly identify who has a stake and who does not. You might want to consider the geographical or sectoral scope of the project (e.g. are you interested only in stakeholders at a local level, or is this a national issue that may involve national or international stakeholders?). Which sectors of the economy or population are relevant to the research? A discussion about these sorts of questions at the start of the workshop should clarify any differing perceptions amongst the group to avoid confusion later.

    b) Choose a well-known stakeholder organisation and **run through the publics/stakeholder analysis for this organisation as an example**. Draw copies of the extendable matrix in Part 4 on flip-chart paper and stick to walls, so that everyone can see what is being done. Explain that interest and influence can be both positive and negative (e.g. a group's interests might be negatively affected and they may have influence to block as well as to facilitate).

    c) Ask participants to **identify organisations, groups or individuals that are particularly interested and/or influential**, and list them in the first column of the matrix in Part 4. I've provided you with a blank table and a worked example to illustrate how this might look. Use the questions in the box below as prompts to help you identify as many stakeholders as possible.

    d) As a group, **work through each of the columns in the matrix**, one stakeholder at a time, discussing the nature of their interest and reasons for their influence etc., and capturing the discussion as best as possible in the matrix (getting participants to capture points on Post-it notes where necessary to avoid taking too long).

e) Take a break, and then invite participants to use the remaining time working individually to **complete the columns for all the remaining stakeholders**, adding rows for less interested and influential stakeholders as they go. Remind people to try and identify groups who might typically be marginalised or disadvantaged, but who still have a strong interest in the research.

f) **Ask participants to check** the work done by other participants, adding their own comments with Post-it notes where they disagree or don't understand.

g) **Facilitate a discussion of key points** people feel should be discussed as a group about stakeholders where there is particular disagreement or confusion and resolve these where possible (accepting differing views where it is not possible to resolve differences).

h) **Identify key individuals to check findings with after the workshop.** Identify up to five individuals from particularly influential organisations, trying to get as wide a spread of different interests as possible (to do this, it may be necessary to start with a longer list and then identify people who are likely to provide similar views to reduce the length of the list). Finally, consider if there are any particularly important stakeholders who have high levels of interest but low influence who you do not want to marginalise and go through the same process to arrive at a list of around seven to eight individuals who you can check the findings of the workshop with.

Box 7 provides some questions and example categories that may help you identify stakeholders. Download the publics/stakeholder analysis tool in Table 4 from this chapter from my website at: www.fasttrackimpact.com/resources. Feel free to change the titles of the columns, as I've done in the worked examples at the end of the book. Just make sure you still capture the interest and influence of each stakeholder, and be aware that the more columns you have, the longer it will take to complete.

Box 7: Useful prompts to help identify stakeholders

A number of questions may be asked during workshops and interviews to identify stakeholders, for example:

- Who will be affected by the research?
- Will the impacts be local, national or international?
- Who has the power to influence the outcomes of the research?
- Who are potential allies and opponents?
- What coalitions might build around the issues being researched?
- Are there people whose voices or interests in the issue may not be heard?
- Who will be responsible for managing the outcome?
- Who can facilitate or impede the outcome through their participation, non-participation or opposition?
- Who can contribute financial or technical resources towards the research?

Example stakeholder categories include:

- Government departments and politicians
- Government agencies
- Industry/producer representative bodies/associations
- Media
- Trading partners
- Land owners and managers
- Special interest/lobby groups
- National representative and advisory groups
- Research organisations
- Professional groups and their representative bodies
- Representative groups e.g. for consumers or patients
- Non-Governmental Organisation (NGOs)
- Community groups
- Local history groups
- Museum curators and staff
- Schools

Chapter 15

How to design events with stakeholders and members of the public

The chances are that your pursuit of impact is likely to involve talking to more than one stakeholder at a time, and that these individuals may have quite differing perspectives. For many researchers, the prospect of having to negotiate and potentially mediate between conflicting parties is their worst nightmare. The good news is that even with the most challenging of groups, you can almost completely design conflict (and boredom) out of your meeting. There is no substitute for working with a professional facilitator to design and facilitate your workshop, but if you don't have the budget or time to hire someone, these suggestions will go a long way towards helping you design an event that delivers what everyone wants and is efficient and enjoyable.

A conceptual model for designing your event

The GROW model comes from the coaching literature and offers a useful conceptual framework within which to think about planning events. It suggests that we need to start by considering the goals of the event, then consider how far the current situation is from the goals you want to achieve, before considering options to get you from where you are now to your goal, and deciding on actions. Although this may sound like common sense, the questions in Box 8 can be a powerful way of checking that your event is action-orientated and contributes towards the goals of your research.

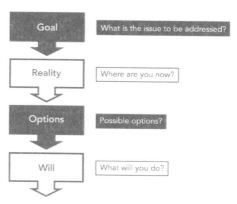

The 'GROW' Model

Goal — What is the issue to be addressed?

Reality — Where are you now?

Options — Possible options?

Will — What will you do?

Box 8: Structuring a process, event or group conversation with GROW

First, think about the goals you have set for working with stakeholders and likely users of your research:
- What do you want to achieve together or change?
- How will you know if you've been successful?
- When do you want to have achieved your goal by?

Next, consider your current reality:
- What stage are you at in your research?
- What are you achieving at present in your research in relation to your goals?
- What action have you taken so far to try and reach your goals? What were the effects of this action?

Next, consider your options:
- What actions could you take to move forward?
- What strategies have worked before in similar circumstances?
- If no barriers or limitations existed, what would you do?
- Which step will give the best result?
- Advantages/disadvantages of this step?
- Which option will you work on first?

Finally, consider what you will do now, at the end of this workshop or meeting with stakeholders:
- What are you going to do?
- When are you going to do it?
- What help do you need?
- Who will you involve?
- What might prevent you from taking this step?
- How can you overcome this?

You can also use this model to structure the overall process within which your event will sit (e.g. a series of meetings and events or activities), and it can be used to structure open discussion during events to ensure it is action orientated and not a talking shop.

Process design

Before considering how to design a specific event, it is important to consider the context in which that event sits. There are two elements to this: the context in which your publics/stakeholders are operating; and your research context. If you have followed the second step in this book, based on the second principle (represent), you should know who is likely to be interested in your research, and what their interests are. You can then ask the following questions to help you design a process that helps you achieve impact from your research whilst meeting stakeholder needs:

- What outcomes do you want from the event?
- What are the outcomes that publics/stakeholders and likely users of your research want (based on your stakeholder analysis — see Chapter 14)?
- Where are the areas of overlap and synergy between your goals and the goals that you think stakeholders are likely to bring to your process? Can you emphasise and focus primarily on these?
- Are there any outcomes you want that stakeholders are likely to oppose, or that stakeholders want and you would not feel comfortable with or able to help deliver? Can you design additional meetings and workshops to negotiate goals with key stakeholders to avoid these clashing interests?
- How does your planned event link to the wider research project, and your funder's and organisation's goals? Can you combine or link your event with another event to make your process more efficient?
- How will you attract people to engage with your event?
- How will you keep people engaged with your research after your event?
- What steps will you need to put in place after your event to ensure you achieve your intended impacts?

Armed with the answers to these questions, you can now develop a process plan in which you organise a range of meetings, events or other activities around your event to ensure you achieve the impacts you want. Decide how many events of which type you need with which groups of stakeholders, and integrate this with your impact plan.

Event design

If you want an event to run smoothly, there are a large number of things you need to do beforehand. There are many important practicalities that are frequently overlooked by researchers when designing events. All it takes is for your venue to tell you that you're not allowed to stick anything up on the walls (as has happened to me on a number of occasions) and suddenly your event plan is in tatters if all your activities involved people writing on posters on the wall. So pay attention to these practicalities to avoid last minute stress:

- How many people do you expect to attend your event? Is your room sufficiently large to accommodate everyone, with extra room for people to move around to do group activities or contribute to material being developed on the walls of the room?
- With larger groups, it can be useful to split into smaller groups for certain activities to ensure everyone has a chance to discuss issues in depth:
  - Do you need to book break-out rooms or will the room be large enough for small groups to be able to work separately around the room without disturbing each other?
  - Do you want small groups to be facilitated or self-facilitating? Getting groups to nominate a facilitator to help steer discussion and capture notes may be efficient, but if they are facilitating properly, it means that you're unable to fully capture the views of that member of the group. Often, naturally more dominant group members may offer to facilitate and then abuse this position by not allowing others to talk or not fully capturing their points in the notes that are developed. This can lead to frustration amongst group members and biased outcomes. Therefore, although more costly and time-consuming, it may be worth assigning a facilitator to each group. Alternatively, to reduce costs, you can approach individuals you think

might be effective facilitators in advance and ask them to arrive early to get guidance on good practice.

- For research projects operating in controversial areas or where there is conflict between stakeholders, you may need to take care to ensure the venue is considered 'neutral' territory. For example, don't accept a free room from a controversial organisation on one side of a conflict.
- Consider how your choice of venue might influence power dynamics within the group you are inviting, for example, might hosting your event at the university intimidate some participants and increase discrepancies in power between those with more or less formal educational status?

- If you are planning to use facilitation techniques that involve putting flip-chart paper on walls, ensure that you have sought permission to do this, as some venues forbid you from sticking things on the walls. Even if you think a flip-chart stand will be sufficient, it is often useful to have the flexibility to be able to put things on the wall so participants can see a record of what has been discussed so far, and build on it in subsequent tasks.
- Is the venue able to provide lunch to participants in a timely manner? Booking a sit-down lunch can lead to unexpected delays, extending your lunch break and taking up valuable workshop time. A buffet lunch may give you the option to reduce time for the lunch break and act as a useful buffer if you're running behind schedule.
- Is the venue fully accessible to everyone you've invited — consider both distance and other accessibility issues, such as whether it is accessible by wheelchair and public transport.
- Have you booked your event at an appropriate time for your target audience? Weekdays will be better for some types of participant, while evenings or weekends may be better for others — you may have to devise two similar events to reach different audiences. Consider the time of year you've booked your event — might winter weather prevent some people from reaching you if you choose a remote location? Are there other key events happening the same day? Is it a particularly busy time of year for some of the professions you're targeting (tax returns due or farmers in lambing season)?
- Do you have all the equipment you're likely to need to carry out your facilitation plan (see below for more information about how to develop an effective facilitation plan)? Even if it's not part of your facilitation plan, it can be useful to travel with Post-it notes and sticky dots, in case you need to give everyone the opportunity to write down their thoughts on a particular issue, or if you need to rank or prioritise anything by getting people to stick dots next to ideas they prefer (more anonymous and easier to record than voting).

Developing an event (facilitation) plan

A facilitation plan is a bit like a detailed recipe for your workshop, which should be self-explanatory and easy to understand for everyone who is helping you facilitate (including you when you're stressed!). Although this may be based around an agenda with timings that match the items on the participants' agenda, it will need to be significantly expanded to provide more details to help you manage the day. In a good facilitation plan you should:

- **Assign a time-keeper** from the team to keep an eye on timings and remind others in your facilitation team when it is time to move on. Provide detailed timings for each agenda item — if you need to do a number of activities to achieve a particular agenda item, list each of these activities and estimate timings. Consider removing timings (or keeping them to a minimum) on the participants' agenda to avoid people noticing if you're running late, so you can easily adapt the programme to catch up time without people worrying they'll be going home late or missing lunch.
- **Assign members of your facilitation team to each activity** in your facilitation plan. Where possible, include a lead and a support facilitator — the support facilitator can help record points, get extra materials when they run out and generally help keep everything running smoothly so that the lead facilitator can focus on the participants.
- **Set clear aims** for your event, and then tailor your techniques to the aims and the interests/needs of participants. For details of techniques you may wish to choose from, keep reading.
- **Make time for introductions** at the start of your event (unless the group size is too large for this) and create time at the end of the day after participants have left for your facilitation team to debrief.
- To ensure your event leads to some practical outcomes, it is worth programming in an **action planning** session at the end of your event where you identify actions that have arisen as a result of your workshop, so you can assign deadlines and responsibilities and follow these up later.
- It can be useful to **start your event with 'opening out and exploring' techniques, followed by 'analysing' and then 'closing down and deciding' techniques** to structure your dialogue as inclusively as possible towards a practical outcome (Box 9).
- It is useful to **include a 'buffer' session in your timings**, such as a long lunch that can be cut short if necessary, or a session that could be cut out if time is running short. This will prevent people feeling rushed, and allow you to spend enough time on the important aspects of the workshop. I usually identify a

session in the afternoon that could be shortened or completely removed without significantly compromising the workshop, in case I'm running short of time or need to create time for a new session in response to a problem.

- **Create an equipment list**, making sure you have all the equipment you need for every activity (don't assume the venue will have anything you can use to stick paper on walls).
- **Trial and test your methods.** If you've not tried a particular facilitation technique/method before, it's never a good idea to try things out for the first time with stakeholders — use it in a research meeting or in class with students first to check you know how it works properly and adapt it accordingly.

Engagement techniques

There are many techniques available to facilitate two-way engagement between researchers and stakeholders as part of the research process (Box 9). I typically start a workshop with opening up and exploratory techniques, before moving on to analysing and deciding techniques. However, you may want to have a separate workshop at the start of your research that is focused entirely on opening up and exploring to understand the research priorities of your stakeholders and adapt your research accordingly. Below I've listed some of the techniques I use most often in my own research. Having these in your mind can be incredibly useful if a technique isn't working for some reason and you need a plan B.

Box 9: Types of engagement technique

- Opening up and exploring dialogue and gathering information with stakeholders about issues linked to your research (goals in the GROW model – see page 169)
- Analysing issues in greater depth with stakeholders, getting feedback on preliminary findings (reality and then options in the GROW model)
- Closing down and deciding on options and actions based on research findings (will in the GROW model)

Opening up and exploratory techniques:

- **Brainstorming** techniques can help rapidly identify initial ideas from a group. By getting participants to think quickly and express their ideas in short phrases, the technique encourages participants to suspend the normal criteria they would use to filter out ideas that may not appear immediately relevant or acceptable. As such, many of the ideas may not be useable, but there may be a number of new and creative ideas that would not have been expressed otherwise, which can be further developed later in an event.
- In a **metaplan,** participants are given a fixed number of Post-it notes (usually between two and five depending on the size of the group, with fewer Post-its being given out in larger groups) and are asked to write one idea per Post-it. Participants then take their Post-its and place them on flip-chart paper on the wall, grouping identical, similar or linked ideas together. The facilitator then summarises each group, checks the participants are happy with the grouping (making changes where necessary) and circles and names each group. In the space of 10 minutes it is possible for everyone to have given their views and you have a summary of the key issues that can be used to structure other group activities.
- **Venn diagrams** can be used for a similar purpose, helping participants identify key issues, and overlaps or connections between them.
- There are a variety of ways to get participants to **list** ideas or information, for example, via responses to requests for information on social media platforms or online discussion boards, or in group work by creating 'stations' around the room where participants can list information or ideas on a particular topic. Stations may, for example, be based around themes that emerged from a brainstorm or metaplan (above). These groups may be facilitated or all participants may simply approach each station and contribute individually in their own time.
- In the **carousel** technique, participants are assigned to groups (with the same number of groups as there are stations) and given a fixed time to contribute to one station before being rotated on to the next. If each group is given its own coloured pen, it is possible for participants to see which ideas were contributed by previous groups. When a group reaches a new station, they are given time to read the contributions of the

previous group(s) or these are briefly summarised by the station's facilitator. They can then query or build upon previous work, listing their own ideas beneath the ideas expressed by previous groups. As the activity continues, it becomes increasingly difficult for groups to add new points, so the time per station can be decreased. Finally, to reduce the time that might otherwise be taken for stations to 'report back' to the wider group, participants can be directed back to their original station to read what other groups have added to their points. Although not fully comprehensive, this gives everyone a good idea of what has been contributed to all stations. For those who want a fuller picture, the materials can be left on the walls to be viewed during subsequent breaks.

Analysing techniques that enable stakeholders to critically evaluate ideas with you include, for example:

- **Categorisation** techniques where participants are asked to sort or group ideas into themes, based on pre-set criteria or based on similarity, for example, the grouping stage of a metaplan, or putting ideas on cards and asking participants to sort the cards into different piles on the basis of their categorisation
- **Mind-mapping** techniques (also known as concept mapping, spray diagrams, and spider diagrams) can be a useful way to quickly capture and link ideas with stakeholders.
- **Problem tree analysis** (also known as cause-effect mapping) is similar to mind-mapping but is a simpler tool (which is also more limited in the way it can be used). It may be useful in settings where the complexity of a mind-map may be considered intimidating for some participants, or where you purposely want to keep the analysis simple and brief. Rather than looking at how all issues are linked to one another, problem tree analysis uses the metaphor of a tree to help visualise links between the root causes and solutions to a problem. A simple picture of a tree is drawn on a large piece of paper, with the problem written on the tree trunk. Participants are then asked to draw roots, writing the root causes of the problem along each root. Some root causes may lead to other root causes, so an element of linking may be done between roots, but this should not get too complex. All these roots lead to the bottom of the tree trunk and at the top of the trunk, branches are drawn, along which potential solutions are written

(again with the potential to link branches to other branches to show how one solution may be **dependent** upon another solution being first implemented). If you want, you can cut out circles of coloured paper to signify fruit, which can be used to represent anticipated impacts or outcomes of implementing solutions.

- **SWOT analysis** encourages people to think systematically about the strengths, weaknesses, opportunities and threats as they pertain to the issues being researched.
- For issues that have a strong temporal dimension or for project planning with stakeholders, **timelines** can be used to help structure discussion in relation to historical or planned/hoped for future events. There are various ways to do this, for example, flip-chart paper may be placed end-to-end along a wall with a horizontal line along the middle of the paper, marking 'NOW' and specific years and/or historic or known future events to help people orientate themselves along the timeline. Participants may then write comments or stick Post-it notes at various points in the past or future, vertically stacking ideas that occur at the same time.

Closing down and deciding techniques:

- **Prioritisation** differs from ranking by enabling participants to express the strength of their feeling towards a particular option rather than simply saying "yes" or "no" (as in voting) or ranking an idea as better or worse than another idea. Prioritisation exercises also enable you to identify options that are considered to be particularly popular (or not) by participants, which you may then want to explore in greater detail. In prioritisation exercises, participants are given some form of counter that they can assign to different options (e.g. sticky dots or, if working outside, stones, but if you don't have anything to hand, people can simply be asked to assign crosses with pens to options). Normally, participants would each be given a fixed number of counters (at a minimum this should be the same number as the number of options) — this prevents certain participants assigning more counters than other participants to the options they prefer, biasing the outcome. If using sticky dots, it is possible to get people to assign different coloured dots to express their preferences according to different criteria (e.g. use red dots to say how cost-effective they think an idea would be and green dots to express how easily they think the idea would work). It is then possible to see at a glance which ideas are preferred, and it is relatively quick and easy to total the number of counters assigned to all options, and if desired, create a ranked list.

- **Multi-Criteria Evaluation** (also known as Multi-Criteria Analysis or Multi-Criteria Decision Modelling) is a decision-support tool for exploring issues and making decisions that involve multiple dimensions or criteria. It allows economic, social and environmental criteria, including competing priorities, to be systematically evaluated by groups of people. Both quantitative and qualitative data can be incorporated to understand the relative value placed on different dimensions of decision options. Broadly, the process involves context or problem definition, representation of evaluation criteria and management options, and evaluation. When applied in a participatory manner with stakeholders, this may involve any of a number of discrete stages, for example:
  - Establishing context and identifying participants: stakeholder mapping/analysis techniques may be used to systematically consider which stakeholders should be involved in the multi-criteria evaluation
  - Defining criteria: criteria are defined that capture stakeholders' interests via facilitated discussion and literature
  - Defining the options that the group is choosing between
  - Scoring options against criteria: the likely performance of each option is scored against each criterion
  - Multi-criteria evaluation: algorithms are used to combine scores and ranks into a weighted value that describes the overall preference towards each option. This may be done either using free software or by hand, adding up scores assigned to each option, and then multiplying scores by agreed amounts for certain criteria (e.g. by 1.5 or 2 depending on whether they are considered to be slightly or much more important than other criteria) and recalculating the scores for each option
  - Discussing the results: this is a decision-support tool so outcomes may be deliberated with participants or amongst decision-makers to assess the degree of consensus, negotiate compromise and manage trade-offs.

I've focused on prioritisation methods in this last section because alternatives like voting and ranking can be problematic in my experience. In most group settings, it can be difficult to ensure anonymity in voting, which may bias results, and there is little room to explore reasons for people's voting preferences. Alternatively, ideas can be ranked. However, getting consensus amongst participants for a particular ranking can be challenging, although the discussions that this stimulates may be revealing. It is also not possible to differentiate between options that are particularly popular or unpopular — this may be important if only one or a few ideas are considered viable, as a ranking may imply that mid-ranked options are viable or somewhat preferred.

At the end of your event you will be left with a mountain of flip-chart paper and Post-it notes. It is always a good idea to photograph everything before you remove it from the walls, in case things get lost or damaged in transit back to your office. I often put sticky tape across flip-chart paper that people have stuck Post-it notes on, to avoid finding a pile of Post-it notes at the bottom of your bag, disconnected from the paper they had been linked to. Be careful to label your folded bits of flip-chart paper so you know which session in your workshop they came from, so it is easier to write up later. Deciphering handwriting and typing this all up yourself can be very time-consuming, so I usually try and get my virtual assistant to do this for me (see Chapter 11). It is important to try and get a report sent to participants as soon as possible after the workshop, even if you don't have time to write much around the tables and photographs that capture the outcomes of the workshop. Make sure you send an accompanying note to anyone who has committed to an action at the end of your workshop. If you don't do this, then there is a danger that people will feel like they have been at a 'talking shop' and it may become hard to re-engage with these people in future work.

Chapter 16

How to facilitate events with stakeholders and members of the public

An experienced professional facilitator is worth their weight in gold. You could run the same event with the same participants, using different facilitators, and get significantly different outcomes. Many researchers think that because they can chair a meeting with other researchers, they can facilitate workshops with stakeholders. This is rarely the case. You will very often be working with very diverse groups with different perceptions of your research, different levels of education and potentially conflicting views. Trying to run a workshop with stakeholders in the same way you would chair a meeting with researchers will rarely get the best out of everyone. In the worst-case scenario, you may end up inflaming conflict and creating long-term difficulties for those you want to work with.

One of the first stakeholder workshops I was charged with designing went very badly when I got the facilitation wrong. It was the first workshop in a funded project that was meant to scope out the potential to conduct a wider research project. The first mistake I made was to ask for the facilitator's day rate when I put the proposal together. When I called her up to engage her for the work, she explained that a one-day workshop involved at least three days of preparation and post-workshop work, so I couldn't afford her. One of my colleagues came to the rescue, recommending an American colleague of his who regularly facilitated stakeholder workshops. For the price of a ticket to a conference, he was happy to facilitate the workshop.

Two things went wrong at the very start. First, our American colleague decided to do a practice run of his conference talk to open the workshop. This might have worked if his talk had something to do with the topic of the workshop, but I could see people shifting uneasily in their seats, wondering if they were at the wrong event. The other thing that was wrong, was that there were three additional people in the room, who I hadn't invited, and I made the mistake of not asking anything about them. Eventually, the workshop started, and people started wheeling out all the old arguments that they'd had for years. The facilitator then stood and watched, saying nothing, as people started raising their voices and being rude to each other. The break-time came and went, and the argument intensified, with the facilitator looking on with a thoughtful expression on his face. At that point I decided that, despite just being a PhD student with no experience of facilitation, I had to put a stop to this. So, I called time on the arguing and we went to the break. I asked the facilitator why he wasn't facilitating, and he explained that he was American, and everyone was speaking in thick Yorkshire accents, and he couldn't understand a word anyone was saying! So, after the break, we moved to a part of the workshop that involved writing things on Post-it notes and sticking them on the wall. However, there was a problem. The three people I hadn't invited weren't doing the exercise. I went and explained it to them, and still they didn't do anything. By now, everyone else had completed the task, apart from these three, and all eyes were on me as I explained the task one last time and asked if they understood. They said that they understood. So, I asked why they weren't doing it. To my shame, they explained that they were illiterate. I wanted to ground to swallow me up at that moment. I realised that I had humiliated them in front of the very people they wanted to influence in this debate, and I felt horrendous. I announced that we would take an early lunch-break, and asked my facilitator if he had any techniques we could use that didn't involve speaking, reading or writing, to which, of course, the answer was "no". Clearly, this wasn't entirely the fault of the facilitator — I had set him up to fail. But it does illustrate how badly awry things can go when the facilitation goes wrong.

Facilitating dialogue with stakeholders and likely users of research

There are a number of reasons why hiring a professional facilitator (or getting a few facilitation skills of your own) can be particularly useful when engaging with stakeholders and likely users of your research during events, for example:

- Efficiency: more can be discussed in less time
- Impartiality
- Clarity
- A helpful atmosphere
- Appropriate techniques

- More people have a say
- No organisation or individual is in control or has the power of veto
- The outcome is open and more likely to be considered fair by all those involved.

Professional facilitation can be expensive, ranging from around £700 to £3000 for a small event, and up to £8000 for a full-day event with over 100 participants. Prices vary according to the expertise/reputation of the facilitator and the amount of time necessary to prepare for the event. Unless their role is little more than that of a chairperson to help you steer your way through a simple agenda on time, you are likely to need a number of days of time discussing your aims and coming up with draft facilitation plans that use different techniques to reach these aims. If you want the facilitator to be responsible for writing up the outputs from your event, then this will cost more. It is therefore advisable to build facilitation costs into your research proposal from the outset.

In many projects, there are not sufficient funds to hire a professional facilitator, so we may end up in this role as researchers. When faced with facilitating an event, most of us are understandably nervous.

Some challenges will emerge from the group itself:

- Dominating people with big egos can be hard to manage. You need to learn techniques for keeping these people in check without upsetting them, so that others have a chance to have their say, and feel able to express themselves freely.
- Equally, quiet or unconfident people can be hard to manage. You need to find ways of enabling them to contribute to the group without putting people on the spot or intimidating them.
- Diverse groups are particularly hard to manage. Groups may be diverse in many different ways, including a mix of quiet and dominant individuals, those with greater or lesser formal educational attainment, those with different levels of power and influence, varying levels of interest in the subject (who are more or less informed about it), and people in a group with very different fundamental values and beliefs.

In addition to this, most of us face a number of internal challenges to becoming an effective facilitator. First, we may lack confidence in ourselves. This may be borne of a lack of experience facilitating events with stakeholders, or it may be a deeper-held lack of confidence that we find emerges in all sorts of public situations where we feel others are judging our performance. Whatever the source of this lack of confidence, there are a number of things that can help reduce your nerves, for example:

- Getting practice: although it may not be possible to practise working with stakeholders, there may be other contexts in which we can try out our facilitation tools and skills, for example, by adapting our teaching with students to incorporate tools and skills we know we'll need to use with stakeholders
- Building in buffer time to your facilitation plan (e.g. sessions you can drop or breaks you can shorten), so you're not creating unrealistic expectations from your event, can help reduce nerves on the day
- Having a facilitation team you can trust to come to your rescue if things seem to be going wrong
- Getting to the venue early so you can sort out any practical issues in good time before participants arrive
- Getting feedback from colleagues on your facilitation plan to make sure it is realistic
- Meeting your facilitation team the day before or in good time before your event to go through the facilitation plan and make sure everyone knows what they are doing
- Considering meeting separately, one-to-one, with any individual you know to be particularly problematic (e.g. argumentative, confrontational), rather than inviting them to the event
- Having a plan B for high-risk activities you have not tried out before can also help reduce your nerves both before and during an event — if a technique isn't working, you know you can change tack. There are also a number of practical tips you can use to keep control of dominating individuals and get the most out of more reticent members of the group (see below).

With practice, there are a number of interpersonal and practical skills that can help you become an effective facilitator. Many of the practical skills are quick and easy to learn, and can make a considerable difference to your practice. However, many of the interpersonal skills are harder to gain. Although some would argue that many of these characteristics are innate and therefore not possible to develop, it may be possible to make efforts to cultivate these characteristics as part of your role as facilitator, though this will take significant time and practice.

It is worth mentioning that interpersonal communication skills are often very culturally specific (though some non-verbal communication transcends cultural differences), so, if you have people from different countries attending, it might be good to know the cultural nuances of those cultures before you go into the room. For example, one of my PhD students, Steven Vella, told me how he once had to jump onto a table and whistle to get the attention of angry stakeholders during a workshop in Malta, threatening to throw everyone out unless they became quiet and asking a member of the project team to apologise for calling them "ignorant locals". This was appropriate in that particular setting, but might have been inappropriate in a UK town hall.

Such interpersonal characteristics of an effective facilitator include, for example, being:

- Perceived as impartial, open to multiple perspectives and approachable
- Capable of building rapport with the group and maintaining positive group dynamics
- Able to handle dominating or offensive individuals
- Able to encourage participants to question assumptions and re-evaluate entrenched positions
- Able to get the most out of reticent individuals
- Humble and open to feedback

Practical facilitation skills include, for example:

- Active listening and understanding. This may include non-verbal feedback such as eye contact, nodding, smiling, focused attention and valuing silence

- Verbal feedback such as sounds, short phrases, clarifying details, encouraging/probing (asking for more information) and using open (not closed) questions
- Giving people time to clarify their thoughts
- Summarising, to confirm that you are interpreting them correctly
- Letting people know their opinions are valued, but without implying that you agree or disagree with them
- Helping people go beyond facts to meanings
- Helping people to 'own' their problems, take responsibility for them and think of solutions
- Reframing points where necessary to help people move from a negative stance to discuss a positive way forward. This involves acknowledging what has been said, and then saying this in a different way that is less confrontational or negative, followed by an open question that seeks to get at the heart of the problem
- Involving others in the group in solving the problem
- Giving momentum and energy
- Ensuring everyone has an opportunity for input without feeling intimidated
- Making an impartial record of the discussion
- Writing clearly, managing paper (ideally with the help of an assistant so you can focus on group dynamics)

Ultimately, to be able to manage power dynamics in a group, facilitators need to have a deep source of their own power. It takes confidence to deal with powerful individuals who are being disrespectful to others in the group. But I'm not just talking about confidence here. It is that thing that you notice in some people, which you can't really put into words; a quiet presence that demands your attention. We have all been in situations where someone walks into the room and you realise that the atmosphere has changed; the conversation might die down and you notice that everyone is waiting for that one person to speak. It is this quiet power that enables the best facilitators to get the most out of the most challenging groups. I would argue that this sort of 'presence' isn't something you are born with, but is something that can be cultivated with commitment and practice.

In Box 10 you'll find a series of questions I've adapted over the years, which are designed to help you understand how powerful you are as an individual in any given context. The answers you give will differ depending on the context in which you ask the questions, so think specifically of a context in which you would like to have more 'presence', so that you can achieve greater impact, and answer these questions specifically in relation to that context. For example, you might ask how powerful you are in the context of your research team or a group of stakeholders (such as healthcare professionals or conservationists) that you need to be able to work with intensively to achieve impact. The first types of power (hierarchical and social) are fairly hard to do anything about, though promotion might come along once in a while. When doing research in Africa, I found that my race and gender were barriers to working with stakeholders in certain contexts. Simply being aware of the power or powerlessness you are likely to feel in certain contexts may help you avoid trying to facilitate in those situations. However, you can work on your personal and transpersonal power. It takes time and commitment to change these ways of being into habits and eventually into characteristics, but it is possible. When I was Director of the Aberdeen Centre for Environmental Sustainability, I knew that I wasn't the most powerful person in the organisation. It was a PhD student. Since I had joined the organisation, I noticed that whenever she had an idea, people followed, and things happened. Despite being at the bottom of the hierarchy, what she had that I lacked was bucketloads of personal and transpersonal power. Her life's goal was to make the world a better place and she had enthusiasm and positivity that was infectious and an altruistic vision that inspired hope. Ana ended up working with me as a Post-Doctoral Research Assistant and together we launched the training programme that this book is based on.

Once you've considered the points in Box 10, it can be useful to share your scores with someone you know well. Discuss which categories you score highest in (e.g. mostly 4 and 5 scores). Where you have low power, can you use higher power from a different area to help you in your interactions with others? Where could you increase your power? Would the person you're discussing this with have scored you differently? If so, why?

Box 10: Identify your levels of power

The following points are designed to help you identify the different types of power you possess in any given context. You can use this in a general sense (thinking about the main social group you belong to or interact with most), but it is most useful to think about how powerful you are in a specific context, for example, as a facilitator leading a workshop with people who are interested in your research. Imagine yourself in this situation, and rate how powerful you feel on a scale of 1–5 in relation to each of the following personal characteristics. You may do this in relation to how powerful you feel and/or how powerful you think the other people in this situation think you are (you will need to choose which of these you think most affects your ability to achieve impact).

Hierarchical power:

- Seniority in formal hierarchy
- Expertise
- Access to decision-makers

Social power:

- Race or ethnicity
- Age
- Gender
- Class or wealth
- Education level
- Strength and breadth of your social networks
- Title (e.g. Mrs, Dr or Prof)

Personal power:

- Self-awareness
- Self-confidence and assertiveness (not over-confidence)
- Charisma and strength of character
- Ability to empathise with others
- Life experience and ability to survive adversity
- Ability to communicate and influence others
- Reputation for integrity and honesty
- Creativity
- Honest estimation of your own worth and abilities, being aware of your limitations and weaknesses, whilst focusing on your strengths and abilities
- Being someone who believes in, trusts and builds up others, rather than criticising and gossiping

Transpersonal power:

- Connection to the other; to something larger, more significant and lasting
- Commitment to a positive and clear set of values and beliefs
- Being prepared to challenge the status quo rather than compromise your values
- Ability to overcome or forgive past hurts
- Freedom from fear
- Service to an altruistic vision or cause

Anticipating conflict

Dealing with difficult individuals and situations can be challenging if you've not got a lot of experience as a facilitator. Despite being a professional facilitator myself with experience of facilitating over 50 workshops with stakeholders, I wouldn't consider myself to be particularly experienced. If I've got a workshop that is likely to involve conflict or particularly high stakes, I will always try and pay for a more seasoned facilitator. But sometimes conflict erupts when we least expect it.

If you've already got to the point where people are having angry outbursts and verbally abusing each other, the chances are it's too late to avoid conflict — you're already in it. But if you can spot the early warnings signs, it may be possible to avert conflict. In my experience, most conflicts with stakeholders arise from power imbalances within the group, so simply identifying particularly high- or low-power individuals will alert you to the fact that some form of conflict may be likely.

Here are a few of the signs you can look for to identify people who are (or are perceived by the group or themselves to be) particularly powerful or powerless:

- In some cultures and organisations, the way people dress denotes hierarchical power e.g. managers in universities often wear suits. Check whether those in your group wearing suits are displaying other signs of high power that could be challenging to manage.
- Who does everyone give eye contact to when they speak, and who never gets eye contact? You've probably had that feeling of being invisible when you're in a meeting where everyone else is more powerful than you (the person taking notes in academic meetings gets this feeling on a regular basis). Equally, you probably know how awkward it can feel when people in a group only give eye contact to you, as though there's no one else in the room. If there is someone in the room that the group perceives to be particularly important, you'll notice that at some point during each person's speech (usually at the beginning and the end), they will give that individual eye contact, effectively seeking their approval and hoping to win influence with them.
- Is there someone in the group who regularly speaks over others and cuts others off? Is there someone in the group who rarely gets to the end of what they're saying, and is there someone else who is always heard out? These are other signs of power and powerlessness that you might spot.
- Do you notice that one person's ideas are rarely picked up by the group, perhaps leading to awkward silence or a change of topic? Do you notice that these same ideas may be suggested later on by someone else and be welcomed and discussed actively?
- Who naturally chooses to sit at the head of the table or near the front, and who avoids sitting at the head of the table and chooses to sit at the back?
- Who has a queue of people waiting to speak to them during the break?
- Do some people display particularly confident or nervous/deferential body language?
- Does one person dominate the discussion, offering their opinion on every discussion point?

- Are some people confident enough to give many people in the group eye contact and do others avoid giving people eye contact or only give you eye contact as the facilitator?
- Do some people feel so important that they can check their laptop and phone constantly rather than engaging in discussion with the group?

Any single one of these signs may not mean anything, but if there are a few of these signs pointing to particular individuals, you might start to watch those individuals for signs of conflict, and adapt your facilitation plan to avoid power disparities becoming any more obvious. You have to be careful not to mistake personal traits for signs of power imbalances or conflict (e.g. someone who is naturally shy or prone to colourful outbursts). In some cases, it is possible to resolve this through effective facilitation, for example, politely asking more dominant people to give others space to contribute, or using a device like 'round robin' to give every person in the group a chance to give their opinion (or pass to the next person if they do not feel confident doing this). Usually, the simplest solution if you're not an experienced facilitator is to move into small groups or move away entirely from open group discussion and use a structured elicitation technique, like metaplan, where everyone has the same opportunity to contribute.

Here are a few of the signs to watch out for that might suggest conflict is imminent:

- Are you noticing people closing their body language (e.g. crossing their legs and arms, dropping eye contact etc.)?
- Are people becoming cold, distant, withdrawn (e.g. moving back from the table, giving one word answers etc.)?
- People often dress up insults as jokes to make it socially acceptable for them to attack someone else and to make it hard for others to criticise them for their comment ("I was only joking"). Look to see who is smiling at the joke — and more importantly who is not smiling. If the person the joke is aimed at is colouring up, the chances are they took the joke as an insult. You might be too late to do anything about it first time round, but you need to watch the situation like a hawk and politely stamp on any future 'jokes' if you want to maintain a positive group dynamic.

- Are people becoming increasingly argumentative, disagreeing and/or blaming each other?
- Are people moralising or intellectualising each other?

But for the really early warning signs of conflict, you need to look inside yourself and empathise with the group you're working with. If you can really get in touch with the way that the group is feeling, and become sensitive enough to your own feelings, you will start to detect the earliest glimmer of conflict and be able to watch out for other signs and act promptly. If there's someone in the room who is feeling really uncomfortable, nervous or angry in the group, the chances are they may project those feelings onto you, or that you may detect their feelings through empathy — and you'll start feeling uncomfortable, nervous or angry yourself. Are you experiencing irrational, unaccountable feelings, urges or thoughts, or acting uncharacteristically out of role? It is likely that this is how someone in the group is feeling. The stronger they feel this, and the more people who feel like this, the more likely you are to pick up on it and experience those feelings yourself. In this way, you can pick up on likely conflict well before there are any visible signs, so you can manage the situation and bring back a more positive dynamic into the group before conflict erupts.

Useful techniques for avoiding conflict

Finally, here are some useful tips you can use to avoid conflict and get the most out of facilitating events with stakeholders:

- Set some ground rules: agree them at the outset, and refer back if needed (people are not to talk over one another, everyone's views should be equally respected, no use of offensive language etc.). It may be useful to write these down and place them on the wall for everyone to see. It is typically easy to agree such rules as a group at the outset. They can be particularly useful if someone becomes obstructive or abusive later in the event. If you are unable to keep them in check, you can remind them about the ground rules that the whole group agreed to at the start. Given that they were part of the group that agreed these rules, it is socially quite difficult for them to ignore them, and if they do continue to ignore these rules, you have a clear basis upon which to ask them to leave.

- Any Other Business (or 'parking space'): if you have someone who finds it hard to be concise and in particular if contributions are off-topic, it is possible to create a 'parking space' where you can write these ideas up and park them to discuss later. This technique only works if the group has jointly agreed to the aims of the event at the outset, and if you have the flexibility to create a 15–20 minute session at the end to deal with the points that are parked. By parking less relevant ideas for later, you can keep the discussion focused and on time. Experience suggests that by the end of the event, it will have become clear to all participants that the points that were parked were not relevant and hence the person who suggested them tends to opt to ignore them at this point. Where points are deemed worth covering, you have created time to deal with them, which prevents these points eating into the rest of your time. Also, because it is done at the end of the meeting, participants are usually keen to finish the event and have an incentive to be more concise at that point.
- Open space: if you discover that your aims do not match the aims of some of your participants, this can be difficult to deal with if you want to keep everyone in the room with you and satisfied with the outcomes. A simple technique is to use some of the buffer time you built into your facilitation plan (e.g. a session you can drop or a break you can curtail) to create an 'open space' discussion. Using this approach, the additional topics that participants want to cover are collected (and grouped if there are many points). Participants then have the option to sign up to topics of particular interest to them over the next break (at this point it might become apparent that some of the topics were just the interest of one vocal proponent, as others don't sign up for that group), and then you facilitate small group discussions, recording points and feeding them back to the wider group. If you don't have enough facilitators to do this, you may ask the person who proposed each topic to facilitate their group.
- Empathise with and mirror your group: get a sense of how the group is feeling (e.g. bored, tired or angry) and adapt your approach to their needs. Empathy is about putting yourself in other people's shoes, so you need to connect with their feeling, identifying with it in some way, such as by voicing it or mimicking it via body language (or both). Then you can start to counter feelings that are likely to negatively affect group

dynamics, gradually changing your body language, tone of voice and language to become increasingly open, up-beat and interested. Although this can take significant effort, you will be surprised at how many start to mirror you and begin feeling and acting in more positive ways.

Chapter 17
Driving impact online

A lot of researchers waste a great deal of time on social media. I'm not talking about sharing cat photos (though many of us find this an enjoyable way to waste time, of course). I'm talking about those of us who engage with social media professionally, but without any particular plan or goal. I'm going to suggest in this chapter that if you aren't working with the public and haven't identified any stakeholders in your research who are likely to engage with social media, then there's no point trying to use social media to generate research impact. It won't work.

Of course, there are many other great reasons for engaging with social media professionally: connecting and keeping up with colleagues around the world, for example, during and after conferences; managing to get in touch with inaccessible professors and politicians who don't reply to their emails; being first to hear about funding opportunities and the latest research in your field; the list goes on.

However, I know many academics who have invested incalculable hours writing a weekly blog that virtually no one reads, or who distract themselves with Twitter throughout the day without actually driving any new interest in their research (or getting much useful information). My hope is that by the end of this chapter, you will find out how you can use social media efficiently, so you don't have to spend much time away from your research, but get significant rewards for the time you do invest, because you invest your time on social media strategically.

Can social media deliver research impact?

You might be surprised how many of us use social media in some shape or form on a regular basis. The reason you might be surprised is that you are probably already using technologies that could be classified as social media without realising it.

There are lots of complex academic definitions of social media, but I think you can boil them all down to this:

Public conversations that take place through digital media.

Using this definition means that YouTube is actually a form of social media, because you can reply to a video with a video of your own, and there are often long public conversations about the content of videos in the comments underneath. Blogs are a type of social media (when they work and people comment on them). Wikipedia can be considered a type of social media if you consider the number of people engaging with and editing the content of some entries. Figure 9 lists some of the other platforms that are currently available.

Typically, around 90% of the researchers I train from all career stages use social media in some way on a regular basis. However, the proportion of researchers actively using social media in their research is actually much lower. For the groups I train, the figure is usually between a quarter and half of participants.

Figure 9: Examples of social media platforms

Here are the top reasons cited by researchers I train who already use social media professionally:

1. You can use social media to get feedback on new research ideas, so that you can reframe them to be more relevant to the people who might use your findings.
2. You can get insights into the way that likely users of your research are talking about the topics you're working on — the kind of language they are using and the sorts of things they're most interested in. These sorts of insights can be invaluable when you need to start communicating your findings.
3. You can be the first to find out about news and events related to your research, and you can link your own work to what's happening, making it more likely that your work is picked up and debated.
4. An increasing number of researchers are finding out about funding on social media (particularly Twitter). You can also identify collaborators for grant proposals, who you already trust to be good team players through your online interactions with them. You can find out about funding that you might not have come across through your institution, especially linked to industry, which can help generate impacts from research.
5. You can take part in discussions around academic conferences using conference hashtags that are used to aggregate content relating to that particular event. You can stay in touch with academics you meet at conferences and elsewhere more easily, and have the opportunity to interact with leaders in your research field on different continents who you might not otherwise meet or be able to interact with.

Why don't more researchers use social media in their work?

Given all these benefits, why aren't more researchers using social media professionally? There are a number of good reasons for limited engagement:

- Time is the number one reason researchers tell me they don't use social media professionally. Most researchers struggle to read and reply to their emails let alone read and reply to the volume of material available on social media as part of their work day.

- Others have real concerns about online abuse. Official statistics show that women are more likely to be exposed to abuse on social media than men, but in my experience working with academics, those working on controversial topics are most likely to suffer abuse, whether male or female. If you work in a challenging area, you have to grow a thick skin to talk about your work on social media. Many researchers rightly ask why they should have to grow a thick skin when they can choose not to engage.
- Privacy is an issue for many researchers, who object to the tacit exchange of personal data, which is then sold to advertisers as the price we pay for engaging with many social media platforms. Many researchers don't want to share their personal lives with their colleagues or the wider world, and others worry about their social media accounts being hacked, leading to identity theft and reputational damage.
- Others point out the space limitations of many social media platforms, which curtails the extent to which you can say anything meaningful in academic terms.
- Finally, there are valid concerns that using social media professionally could get people into trouble — ill-chosen words on social media have, after all, cost many people their jobs.

All of these are genuine concerns, and it is important to be aware of these issues before you consider whether or not you should be engaging with social media in your work. If you read this chapter and come to the conclusion that the risks are too high, and that you will not use social media in your research, then I will still have done my job. All I want is for researchers to take a serious look at the risks and benefits, and make an informed decision about whether or not to use these tools. What I'd like to avoid is people deciding not to engage out of fear or because they believe some of the greatest myths about social media for researchers.

The four greatest myths about social media for researchers

Most people who do not use social media in their research do so for good reasons. But I think many of the reasons people give for not engaging with these technologies are in fact myths. I believe that these are the four greatest myths about social media for researchers:

1. *"Productive researchers don't have time to waste on social media"*

To avoid wasting time, you need to be aware of the amount of time you spend on social media, and what you are getting from that time. Do you actually have any idea how many minutes per day you currently spend on social media? How much of that time is spent during your work day, and how much benefit (versus distraction) do you get for your work? For me, the easiest way to understand whether or not social media is a help or a hindrance, and understand if I'm spending too long on it, is to regularly audit the time I spend on social media. In addition to giving you a wake-up call, the process of monitoring your time on social media for a week helps call your relationship with social media to consciousness, so you can think more clearly about how you interact with it. I only use social media on my smart phone, so I can use an app tracker to tell me how long I spend in each social media app per day. The alternative is to download a time-tracking app where you can manually set a clock when you start a task, and stop the clock when you switch to another task. These apps typically colour code different sorts of tasks for you so you can see how long you spend on social media, email and other tasks, and how little time you spend each day on some of the more important parts of your role. It is surprising how much easier it is to resist the temptation to check social media or email when you are holding yourself to account and know you will have to press 'stop' on your reading or writing task in order to 'start' more time on social media.

Rather than just limiting your time on social media, I'm going to suggest that you can go a step further. I have discovered that it is actually possible to save time and reduce the length of my working day through my use of social media. At the moment, I'm saving almost an hour per day, which I've chosen to re-invest in some better work-life balance (by taking out a subscription to Audible and buying a Kindle, and doing some recreational reading). This is how I do it...

Do you read or watch the news most days? If so, then you probably have a news-shaped space in your daily routine. The problem with the mass media that most of us consume, however, is that it is not very targeted. You have to wade through pages of newsprint or listen to or watch a whole broadcast to catch the few items that really interest you. I'm not saying that you should stop engaging with print and broadcast media, but what if you were to cut down the amount of time you spent on that, and filled that time back up again with highly specific news that's particularly relevant to you? Wouldn't this time be better spent? What if, during that news-shaped space in your schedule, you'd heard about the latest discoveries in your field, found out about a grant you could apply for and kept abreast of policy developments or commentary relating to issues you are researching? Would that be time wasted or would that time actually make you more productive? I have found out about funding opportunities and found collaborators for grants (that I've subsequently won) through social media.

What about all that frivolous stuff you hear about on social media all the time? You don't want to watch any more cat videos. Fair enough. Me neither. The great thing about most social media platforms is that you can unfollow or mute the people who are boring you with endless pictures of their pets. I work on my signal-to-noise ratio all the time, unfollowing or muting people whose material isn't relevant enough to be worth my time. The result is a tailored news stream of highly relevant material whenever I've got time to look at it.

I have over 60,000 followers across my social media accounts. You would think I must spend hours on social media every day, but the reality is surprising. According to my latest audit (see the start of this section for my method), I currently spend about 35 minutes per day on social media (most of this time is spent managing two out of my four Twitter accounts and my LinkedIn account). Within this 35 minutes, I get all my news (I now don't use any other source of news media other than Twitter) and pursue pathways to impact for my projects. This is a net saving of 55 minutes per day, compared to the 90 minutes I used to spend on the news. Some of that time engaging with news used to be listening to the radio on car journeys or washing dishes. Now, I find starting or ending my work day with an audiobook much more relaxing and uplifting, and I feel like I have better work-life balance. If you are able to make a net saving in your day by getting your news via Twitter, you can of course choose to reinvest this in work-life balance, as I have, or in getting more done in your work day. The choice is yours.

If you're thinking that's easy for someone at my career stage but not achievable if you are an early career researcher, then take a look at Rosmarie Katrin Neumann (Twitter handle @RosmarieKatrin). She crowdfunded part of her first year as my PhD student, before she was awarded her scholarship at Newcastle University. Over her first four months on Twitter, she attracted 55 followers. Then she started using the techniques I describe in the following chapter. Within the next four months, she reached 1,000 followers, and two years later she had >10,000 followers (including many of the big names in her field). This is not only important for her visibility as a researcher, but also for building networks as she is starting her own business as a knowledge broker alongside her PhD. She needs her ever-growing network to make people aware of the services she provides. She spends between 15 and 30 minutes per day on Twitter, including catching up on news and following/unfollowing people linked to her interests. Anyone can do this, and it doesn't have to displace other work.

I don't want this to come across as suggesting that our goal should always be to amass as many followers as possible. The focus, if we want to generate impact, should be on the quality of engagement you can derive from your use of social media. In many cases, a small but highly engaged and relevant following is far better for achieving this. However, for certain purposes, you may want to become influential on social media, and for that you need to be well known. I have different strategies for different projects. For my research project Twitter accounts, I'm focusing on providing balanced, up-to-date evidence to inform policy and practice. I tweet anywhere between once a week and once a month from these accounts, and don't have a strategy for growing my followers on them. However, for my knowledge exchange and impact research (@fasttrackimpact), I gave myself a target of reaching 10,000 followers before I launched my free online training course for researchers, because I wanted to have enough influence and visibility to be able to make the course widely available. It was an ambitious goal, given that it had taken me three years to amass 2,500 followers, and I only gave myself three months to reach 10,000. Two years later I have almost 50,000 followers on that account and it is the largest social media account in the world solely focused on research impact.

This social media strategy is just one strand of a much wider impact plan designed to get my research on impact used by as many researchers as possible. Social media enables me to achieve specific, measurable knowledge exchange milestones on my pathway to impact (like the follower targets I just described), but ultimately I measure the impact via longitudinal surveys with researchers, six months and a year after I have trained them. As I explain in Chapters 8 and 22, it is as important to track the success of your knowledge exchange as it is to track your impacts. How will you know if you are on track to achieve your impacts if you have no idea that your knowledge exchange activities aren't working? The nice thing about social media is that as a form of knowledge exchange, it is very easy to track your progress.

2. "Social media will intrude on my personal life"

Okay, if you've got an addictive personality, this might not be a myth: proceed with caution. But assuming you can manage the temptation to check your social media networks at every opportunity, I think many people's privacy concerns are very real, but entirely manageable.

First, you don't have to put photos of your breakfast on social media — that choice is entirely yours. You don't even have to post things — you can simply use social media to consume material. Second, you can set most social media platforms to only allow those you want to see your content. Even though I don't have many friends on Facebook, and many are family members, I never post personal stuff, and I've got it set so that if others post personal stuff about me, I get to review it first before it appears on my timeline and the timelines of my friends. Third, you can choose to only use social media from your computer and if you do have it on your smart phone, you can choose to turn off the notifications so they don't intrude on your personal life.

3. "No one would be interested in anything I've got to say anyway"

You don't have to say anything. Most people start their use of social media as 'watchers' — they watch what other people are saying, and use social media purely as a form of news. Many people stay in that mode of engagement, which is still really useful. However, many people then graduate to liking, sharing, reposting or retweeting the things they find most useful (Figure 10). If you stop at this point, that's also great. Now you're not only benefiting from what other people are saying, you're adding value to others like you who are following your updates. Many researchers engage with social media in these two modes for years before they post any of their own material. You don't have to have anything interesting to say to benefit from engaging professionally with social media.

It is worth saying that when I give people the challenge of summarising their research area or a recent finding in 280 characters or less, there is very rarely anyone who can't do it, and the things you learn about people's work from what they've written can be fascinating. Being forced to be concise and simple in our language can be difficult for many academics, but it is surprising how engaging you can be when you try. Even if you really can't find anything particularly interesting to say about your research, there is a very high probability that other researchers in your field will find it very interesting. Although that may not drive impact, it can still help you build your professional networks.

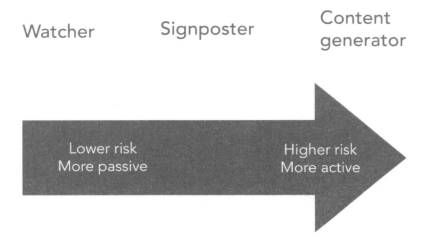

Watcher   Signposter   Content generator

Lower risk
More passive

Higher risk
More active

Figure 10: Most researchers start by watching posts by others, before signposting followers to useful material and eventually generating their own original content. Risks to your time and reputation increase from left to right.

4. "Social media will get me into trouble"

One academic told me that he had banned himself from social media because he couldn't trust himself not to say something he'd regret after a couple of glasses of wine on a Friday night. For most of us though, the chances of something going badly wrong are fairly remote if you exercise a bit of sensible caution. We've all heard about high-profile people losing their jobs over misjudged tweets, but part of the reason that they lose their jobs and we hear about it is because of their profile. You just have to look at the horrendous things that trolls say without consequence to realise that there is a lot of latitude in what people can get away with. However, as researchers, we don't want to be just getting away with it — we have our professional reputation to protect. So, my advice is to be super careful online and remember that everything you say is on the public record. You can say far more on a public stage than you can on social media, because of the way comments online can be taken out of context so easily.

For example, I once arrived late to speak at an event in Windsor Castle (after getting it mixed up with the Tower of London — oops!) and missed the bit where they said that the meeting was being held under Chatham House Rules (where you're not allowed to reveal the identity of those present). I subsequently got into trouble for tweeting a photo that showed who was attending the meeting. So yes, social media *can* get you into trouble, but for most of us, with a bit of care, getting into serious trouble is extremely unlikely.

The most dangerous thing you can do is to dive straight in at the deep end and start posting your own material on a platform you have just joined. You are quite likely to make a mistake, such as posting something private in public. Instead, start slowly and learn the culture of the platforms you are using before engaging in content generation (moving from the left to the right of the arrow in Figure 10). Risks to your time and reputation increase as you move from watcher to signposter to content generator. Even as a watcher, where you only read from the platform, there are risks. For example, it may be possible for someone to see who you are following, and to infer your political beliefs from the political affiliations of the accounts you follow, which may undermine your neutrality as a researcher advising policy. To mitigate this risk, I follow accounts from across the political spectrum from my personal account. Moving towards the middle of the arrow in Figure 10, people are much more likely to infer your opinions from the posts you like or share with others (no matter what you may write in your bio about retweets not being endorsements). Moving to the right of the arrow, when you start generating your own content, you run the risk of being quoted and taken out of context.

If you want to engage with social media without taking unnecessary risks, move slowly from the left to the right of Figure 10:

- **Watcher**: Start by signing up to a social media platform like Twitter or LinkedIn and just connecting with and reading from relevant people and accounts. If you choose who you follow carefully and manage your signal-to-noise ratio by unfollowing less relevant accounts, you can get immediate benefits for your research by efficiently staying on top of the latest developments and funding opportunities in your field. Engaging with social media as a 'watcher' can also prepare you for generating impact. Start connecting with high-level politicians, journalists and industry leaders who might be able to help you disseminate your research and achieve impacts. Many journalists have their mobile phone number in their profile and many leaders will respond to private messages on social media directly despite the fact that you cannot reach them via letter or email. Start following people you think might benefit from your research and listen in to their public conversations and comment, so that you know the language they use and the issues that are resonating with them. When you do meet these people (or people like them) face-to-face, you are much more likely to be prepared for the difficult questions and be able to use language that will resonate.

- **Signposter:** The next step most researchers take is to start signposting people to useful resources online. It may be your latest paper, an article you read via social media that morning or something you're about to send to your PhD students or research group. Now, rather than just sending the email, you are repurposing your email and posting the link to the story or paper on social media. Typically people will just copy or paraphrase the title of the piece they are sharing, so these are not your words that can be taken out of context or used against you.
- **Content generator:** The final step that researchers take, typically (and advisably) after spending a significant time learning the ropes as a watcher and signposter, is to start actually posting their own content based on their research. This is the point at which most opportunities for generating impact occur, but if you're going to invest the time and energy in generating new content, make sure you've got a clear social media strategy so you know that you are using your time wisely.

Finally, it is important to emphasise that you get to decide for yourself if you want to engage in higher risk activities online that are more likely to generate impacts from your research. No one should make you feel left out or like you are a dinosaur because you have decided that you do not want to engage with social media. Weigh up the potential benefits and the risks, and then make a decision you feel happy with and stick to it with the confidence that you have made an informed decision.

How to make your digital footprint work for you

The key to making your digital footprint work is to understand why you have one. What do you want to get out of the time you invest in your online presence? Are you getting these things? If so, are these things worth the amount of time you have to spend online to get them? If not, how can you pull back from your online engagement to stop wasting time? Some researchers have a digital footprint for one reason only: their employer demands it. If there is nothing else you want or need from the online world, then just make sure your institutional profile is up to date and you are contactable, and your job is done. For many researchers, their reason for being online is linked to their research only: access to better information more efficiently, opportunities to collaborate or get funding, and so on. For other researchers, online engagement is a pathway to impact. Depending on your reasons for being online, you will need to invest more or less time in different ways.

Once you have established why you are online, the next step is to take stock of your current digital footprint, so you can assess whether or not you need to make changes. Here are a few quick and easy initial steps you can take:

- **Audit your digital footprint**: do a Google search for your name and the institution you work for and see what comes up. If you've Googled yourself before, it is worth downloading a new browser or using a colleague's device as Google will know that you are looking for you and not someone with a similar name, and automatically rank your institutional profile close to the top of the list. This is not what others searching for your name would see, unless they had searched for you a number of times in the past.
- **Interrogate your online identities**: what profiles come up when you search for your name? Are they for you or someone else? Is your main institutional profile on the first page or do other profiles get listed first? Do these other profiles represent you the way you would like to be seen by the outside world?
- **Prune, cultivate or consolidate your online identities**: first remove any non-professional identities or make them private. Next, ask yourself how each of these different profiles benefited you in the last year. If you aren't getting any value then don't waste your time keeping them up to date — remove your profile and focus your limited time on the profiles that are

most likely to bring you the benefits you are seeking for your research. As part of this, you may consider consolidating many profiles into one or a few that you can more easily keep up to date. This may be as simple as ensuring that you have got links signposting the most relevant profiles (e.g. your Google Scholar publication list and Twitter account) from the profile that comes up first in a Google search (e.g. your institutional profile).

- **Actively manage your digital footprint:** regularly review and update all your online profiles every six months or so.

There are a number of low-risk online platforms for researchers to communicate their research that are worth investigating:

- If you've got an academic email address you can get a **Google Scholar** profile (Figure 11). Google will automatically populate your profile with your publications (you can correct it if there are mistakes) and rank them by citations. Now whenever one of your papers turns up in a Google Scholar search, your name will be hyperlinked from the author list to your profile so people can read more of your work, which could help boost citations.
- Unlike Google Scholar, **ResearchGate** (Figure 11) and **Academia.edu** are actually social media platforms because they enable researchers to engage in debate around the publications they list. Although higher risk than Google Scholar, which does not allow this, the networks are only open to researchers, so risks of online abuse are lower than public social media platforms. These platforms also automatically populate your profile so they don't take a lot of time. You may need to alter the settings in ResearchGate though to prevent it spamming your co-authors on your behalf whenever it finds new papers you've written.

When you join a new platform, make sure you are clear about your reasons for joining, and soon after you have joined, assess whether or not you are getting what you want from it. For example, I joined ResearchGate and Academia.edu at the same time to experiment with both platforms, and then closed one down to reduce the amount of time it took to keep things up to date. The danger is that your digital footprint grows arms and legs as you join new platforms, forget about them and they go out of date. This is an increasing problem for many researchers.

Figure 11: Google Scholar and ResearchGate profiles

Bringing coherence to a fractured digital footprint

The number of digital platforms that profile researchers' work is proliferating rapidly, leading to an increasingly fractured picture of their work. For some researchers, this problem is compounded by the fact that they work on multiple, very different issues, with different communities of researchers and stakeholders. Some researchers are experiencing unintended digital sprawl. Others have chosen to cultivate different identities across different platforms and accounts to engage with specific communities. Either way, the proliferation of online identities can be confusing for people who just want to know who we are and what we do.

As a result, many researchers ask me which one platform they should be on. This is an attractive option in theory, as you only have one place to update. If you have to choose one place, then your employer will probably tell you that you should choose to focus on the profile on your institution's website. That is good enough for many researchers, but many want their work to be more visible or want to have more control over the way they organise and present their work than their institutional profile allows.

If you want to keep digital sprawl under control and are looking for just one other place to feature your research (other than your institutional profile), then your choice will need to reflect what you want your digital profile to do for you. If you want to reach out to other academics and primarily showcase your academic work, then Google Scholar and ResearchGate are popular platforms which make your work highly visible with minimal time input (they identify your publications for you, so you don't have to manually input them). If you want to face a broader audience, or aren't sure if you'll be hanging around in academia for long, then LinkedIn enables you to showcase publications, projects, presentations and more to a broad professional audience, but it requires manual entry.

To help you to decide which platform to focus on, check and see which ones already get ranked highly in a Google search for your name and employer. That way, you guarantee that people looking for you find something relevant and up to date, and you can signpost them from there to a small number of less highly ranked sites. By channelling traffic in this way, you can very quickly bring coherence to a fractured digital footprint.

You don't have to settle for one or two platforms, though; indeed, there are compelling reasons for engaging across multiple platforms, for example, to engage with different audiences via specific social media accounts that enable you to connect with specific groups of people. However, the more different profiles you have, the more confusing the picture may become for those looking in. If your digital profile is spread across multiple websites, platforms and accounts, then you need to find a way of bringing together these fragments to create a more coherent image of your work.

Your first challenge is to find a unifying phrase, concept or strapline that summarises the full range of your work effectively. As a researcher who studies people's interactions with the natural environment and knowledge exchange for impact, I came up with "knowing people, knowing nature" to summarise my work. I am often surprised at how quickly and effectively researchers manage to succinctly summarise what they do in plain English when I ask them to do so in a tweet (280 characters) during social media training. Give it a try...

Now you can copy the same phrase across multiple platforms, making it clear that you are the same person and not a different researcher. You can also start linking accounts, providing hyperlinks to the places you feel best represent your work and which you keep up to date most regularly. This can be a nice way of avoiding spending too much time on sites where you have to manually enter your information (like your institutional profile), instead providing summary information and signposting to sites that are easier to keep up to date. You can do the same with social media accounts (for example, in the biography for @fasttrackimpact on Twitter, it says that "tweets are by @profmarkreed" and in the biography for @profmarkreed, it says that I do "research impact training @fasttrackimpact").

Your final option, which is the most powerful (but also the most challenging to pull off effectively), is to create your own personal website which all other platforms point to as the main source of information about your work. On your personal website you can configure your material in any design you want, and have full control over updates. You can create your own narrative that links your various identities as a researcher, pointing people to specific platforms if they want to engage with you about those issues in greater depth. If you link to your personal website from your university profile, Google should fairly quickly start to rank your website at a similar place in search results to your institutional profile.

Figure 12 shows examples of personal websites made for researchers by Fast Track Impact. The front page of Christopher Raymond's site links to the other platforms he is active on (you can visit it at: www.christophermraymond.com). At the top of Heather Flowe's publications page, she has featured papers that she wants to be read and cited (see: www.heatherdflowe.co.uk). On my own website, you can see the strapline I created to try and sum up the diverse research I do (see: www.profmarkreed.com).

You can contain (or at least make sense of) the digital sprawl, and if you do, you may find that you spend less time updating the 'digital you' and more time benefiting from the collaborations that arise from a coherent digital profile.

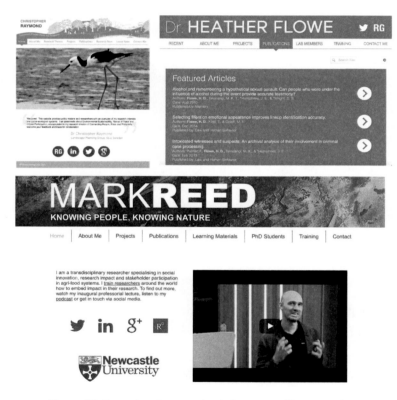

Figure 12: Examples of personal websites created for researchers by Fast Track Impact

How to use social media more strategically to drive impact

When I was asked to apply for my current Chair position, I tried to protest that I was in the middle of my longest ever losing streak for research funding and that I could never live up to the reputation of the retiree who had vacated the Chair (who is my all-time academic hero). I was told that my record spoke for itself, according to my website, and that I should think about it. The majority of plenary talks I'm invited to give come via my website, and I've built an international training business through social media, without ever once paying for advertising. Time invested building and curating your digital footprint really can pay dividends for your career.

Moreover, having a strong digital brand as an academic engenders credibility and trust with many of the people who you might like to use your research, making it easier for you to connect with them. The next section shows you how you can harness the power of social media to drive research impact. If you use social media strategically, as I'm going to suggest, then it doesn't have to take up huge amounts of time. It doesn't have to intrude on your personal life or get you into trouble. And you may discover, to your surprise, that you've actually got some quite interesting things to say, which other people find both engaging and useful.

Researchers are in a unique position on social media because we have easily verifiable credibility as authoritative voices. Box 11 uses a fascinating (though somewhat grisly) example to illustrate how this works. Although it is unlikely that any research finding would ever 'go viral' to the extent that the information in Box 11 did, it is possible for researchers to achieve significant reach via social media. I once did an experiment with a government department to see if we could use Twitter to get public feedback on a policy consultation. The experiment failed because (as you could probably have told me) no one was able to provide meaningful feedback in 140 characters. However, it did raise awareness of the consultation, with some of our tweets reaching a potential audience of over 40,000 people.

Figure 13: Screenshots of Twitter analytics for one month from Twitter.com

An analysis of @fasttrackimpact on Twitter over a period of one month shows that my 20 tweets were seen by 218,000 people (Twitter's definition of 'impressions') (Figure 13; that figure rises to a total of 5 million tweets seen over the last two years). I was mentioned in the tweets of 112 people, 4,937 people viewed my Twitter profile to find out more about Fast Track Impact, and I got 352 new followers that month. Twitter's analytics even tell me the interests of my followers and demographic information that might help me further tailor my messages to my audience. These statistics may not be 'viral', but they illustrate the power of social media to disseminate messages. Karine Nahon and Jeff Hemsley, in their book *Going Viral*, suggest that: "*a viral information event creates a temporally bound, self-organised interest network in which membership is based on an interest in the information content or in belonging to the interest network of others.*"

Believe it or not, as researchers we can create a 'viral information event' based on our research. In fact, as researchers we have an advantage over almost anyone else if we want to be listened to. Since the 1950s, research has shown that people are more likely to adopt the position of a source if they perceive that source to be credible.

People are surprisingly discerning in what they trust and believe on social media, and if you have a link to your institutional webpage and are clearly who you say you are, you have instant credibility in the eyes of many social media users. This means that our voices carry weight in this sphere, and this gives us an immediate head start if we want to communicate our research to a wide audience and engage people in conversations about our work online. Although we might constantly tell our students to check their sources and use peer-reviewed material, the average person, including many decision-makers, rely on anecdotal evidence that they find online. Instead of worrying about this, we can do something about it by adding our voice to the debate and making high-quality evidence accessible to those who are debating the issues we research. To do this, however, we need to go beyond digital dissemination and online marketing to having digital conversations.

Most academics use social media without any clear plan — they're just putting out material and hoping for the best. But if you really want to harness the power of social media to generate impact from your research, you need a plan. If you don't have a plan, then you may well be blogging and tweeting into empty space. It doesn't take long to think strategically about your use of social media, but when you do you'll discover that your time on social media has never been better spent. With a clear plan of who you're trying to reach and why, you can take your use of social media to a completely new level.

Box 11 The role of credibility in viral communication

At 21.45 Eastern Time on 1 May 2011, the US White House announced that President Obama would be addressing the nation in 45 minutes time. Naturally, rumours rapidly began to circulate, as people attempted to guess what the announcement would be about. Two theories began to circulate, namely that either Muammar Gaddafi, former ruler of Libya, or Osama Bin Laden had been caught. However, neither theory gained particular traction until the appearance of the now famous tweet by Keith Urbahn, chief of staff to Donald Rumsfeld, 38 minutes after the news conference was officially announced. It was his position, visible on his Twitter profile, that gave him the credibility to overcome the rumours and initiate one of the best-documented viral events on social media.

Keith Urbahn
@keithurbahn

✿ ☌ Follow

So I'm told by a reputable person they have killed Osama Bin Laden. Hot damn.

RETWEETS 1,669   LIKES 724

3:24 AM · 2 May 2011

Despite only having 1000 followers on Twitter, after one minute there had been 80 reactions (retweets and mentions), and this reached 300 after 2 minutes. In that second minute, New York Times reporter Brian Stelter tweeted, "Chief of staff former defense sec. Rumsfeld, @keithurbahn, tweets: 'I'm told by a reputable person they have killed Osama Bin Laden."

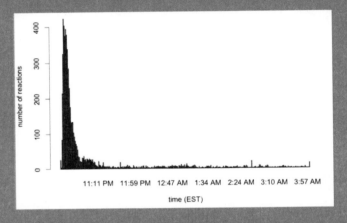

Number of times Keith Urbahn's tweet was retweeted

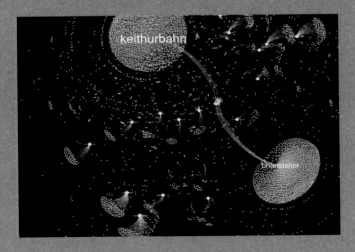

Brian Stelter had 50,000 followers, and his tweet was retweeted hundreds of times in a matter of minutes. Twenty-four minutes after Keith Urbahn's original tweet, the news was being mentioned on Twitter 30,000 times per minute. What is particularly interesting about this example is that many others were guessing correctly before Keith Urbahn's tweet, and rumours about Colonel Gaddafi continued to circulate without ever taking off. The key reason that this tweet initiated a viral information event was the credibility of the source.

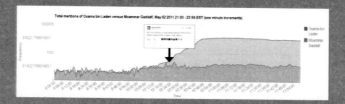

As a researcher, if you link to your institutional webpage and/or mention your university in your Twitter profile, you instantly have credibility in the eyes of the average Twitter user (whether you deserve it or not). This means that your voice can carry weight in discussions on social media.

Images from socialflow.com

How to make a social media strategy

The easiest way to make sure your time on social media really counts is to have a social media strategy. If you can answer these four questions, then you've got yourself a social media strategy. Simple.

1. What offline impacts do you want to achieve via social media?
2. Who are you trying to reach, what are they interested in and what platforms are they on?
3. How can you make your content actionable, shareable and rewarding for those who interact with you so you can start building relationships and move the conversation from social media to real life?
4. Who can you work with to make your use of social media more efficient and effective?

You don't have to write anything down — you just need to act on the answers to these questions to stop wasting time and start generating impacts on social media. If you want to write stuff down, Table 5 is a logic model that breaks each of these questions down and provides you with prompts to help ensure that you are doing things on social media that will credibly lead to real impact (rather than just followers, views or likes). You can download an editable version of the table from my website at: www.fasttrackimpact.com/resources.

1. What impacts do you want from your use of social media?

If you want your time on social media to really count, you need to know exactly what you are trying to achieve:

- Use the Fast Track Impact Planning Template to devise a broad impact plan for your research (Table 3, Chapter 10).
- Identify any strands to your impact that could be achieved via social media, and write your impact goals for social media in the first column of the template. Try and make them as specific, measurable, attainable, relevant, and timely (SMART) as possible.

Table 5: Logic model showing how each of the four questions in a social media strategy can be linked to impact goals to ensure you design a strategy that efficiently drives the impacts you are interested in

Impact goal	What would I expect to see happening offline that would indicate my engagement with social media is moving me closer to this impact goal?	Which stakeholders or publics on social media can help me reach this goal?	Which social media platforms are these stakeholders and publics most active on?	What aspects of my research are these stakeholders and publics most likely to be interested in?	Linked to these interests, what content, resources or opportunities would these groups find most valuable or rewarding?	What actions or activities could I promote via social media to encourage deeper engagement with my research, which might lead to conversations offline that could help achieve impact?	What are the main social media accounts that have content linked to this impact goal? What can I learn from their most popular material? Regularly update this list of accounts and insights, and promote your work to their followers by directly requesting retweets/likes or following their followers.

- Identify indicators that will tell you if your use of social media is taking you closer to the offline goals you have identified from your impact plan. Social media metrics are the easy part. Put some thought into easily-measured indicators that will tell you if you are beginning to translate online influence into offline impact.
- Choose easy-to-measure indicators that will show you whether or not your engagement with social media is moving you closer to your impact goals.

2. Who are you trying to reach?

If you know your audience, you will be able to generate content that they love, and start to build influence online:

- Use the publics/stakeholder analysis template (Table 4, Chapter 14) to systematically identify groups of stakeholders and publics most likely to be interested in your work, or who might benefit from or be able to use your work to help you achieve your impact goals. Write down these groups in second column of the template.
- Using the same publics/stakeholder analysis template, identify what aspects of your research these stakeholders and publics are most likely to be interested in.
- Identify which social media platforms these stakeholders and publics are most active on. If you are using a different platform, you will need to switch.

3. How can you make your content actionable, shareable and rewarding?

If you want your message to travel far and wide, it is worth thinking about how to craft messages that are shareable, rewarding and likely to lead to actions that can help you reach your impact goals:

- What actions or activities could you promote via social media to encourage deeper engagement with your research, which might lead to conversations offline that could help achieve impact? For example, you might be trying to get people to read your blog, comment on it, cite your work in their own blog, sign up for a newsletter, sign a pledge or come to an event.

- What can you do to make your message more shareable? There is evidence that social media messages with images or videos are more likely to get engagement than text-only messages. Is there some way that you can draw a parallel or comparison to your research findings that might surprise people, and make them more likely to share your work?
- Linked to the interests of your stakeholders and publics, what content, resources or opportunities would these groups find particularly valuable or rewarding? Often people follow researchers, research projects or institutions on social media because they are looking for up-to-date, unbiased coverage of the issues they are interested in, so consider how you could offer information that exceeds expectations in terms of both its quality and accessibility. If someone clicks on a link expecting to see the latest research and discovers it is much easier to understand than they expected because the research is described in a blog with an infographic or short video, and they discover lots of other useful resources on your website, they are more likely to tell others about your work and drive traffic to your site for you.

4. Who can help me?

If all this sounds way too time-consuming, then there are a number of ways you can get your message out without having to invest time in building your own following on social media, for example:

- Identify the social media accounts that have content linked to your impact goals and have large followings. Then approach the account owner to ask them to share or repost something you've written from your own account. Alternatively, give them the link and some suggested accompanying text, and ask them to put it out on your behalf from their account. I usually start by asking them via the platform, moving to email and then telephone if I don't get a reply. For large organisations, ask to speak to a member of their press office or social media team, and once you find the person sending out the material, make your pitch to them. If you have targeted an account that is working close to your field, then it should be easy to make the argument that your material will add value to their work.
- Consider taking a team approach to your social media, if there are others you work with who are naturally good at crafting

messages on social media or who already have large followings. Agree to promote each other's material where relevant. Consider asking your post-doc to manage the project social media presence if they are naturally more competent than you.

Case study

I'll conclude this chapter with an example of a social media campaign I helped with, linked to my research. I will explain how we answered each of the four questions above to generate impact.

In 2014, I helped develop a #peatfree campaign to promote the use of peat-free composts instead of peat-based ones, which are leading to the destruction of lowland peat bog habitats around the world. The campaign was led by Project Maya (www.mayaproject.org), supported by a host of minor celebrities and charities, and drew on research I had done with the International Union for the Conservation of Nature's (IUCN) UK Peatland Programme in one of my research projects (Sustainable Uplands).

What offline impacts do you want to achieve via social media? We started with two clear goals. The first was a long-standing goal I had developed with IUCN to restore 2 million hectares of damaged peatbogs by 2025 via a private-public policy mechanism based on my research. This dovetailed with Project Maya's goal to raise public awareness about the hidden value and beauty of peat bogs. I was particularly interested in this goal because we had feedback from companies we had approached to invest in peatland restoration that their customers and stakeholders didn't like peat bogs. Project Maya had their own linked goal, as a Community Interest Company with peat-free products that they were using to raise money for their charitable work, buying and turning inner city land into nature reserves and allotments to give back to local communities. As the work progressed, however, we found that there was strong interest from the global policy community in our campaign, so we added an additional impact, to "inform global policy to generate far-reaching and significant benefits for human well-being, climate and nature through peatland restoration". More specifically, the goal was to facilitate the uptake of the policy mechanism we had designed in the UK in at least one of the countries responsible for 95% of greenhouse gases from global peatlands, leading to new peatland restoration in that country.

Who are you trying to reach, what are they interested in and what platforms are they on? Initially, our target audience was UK gardeners who currently used peat-based composts, but as the work progressed, we also targeted the international policy world. There is a thriving community of gardeners on Twitter, so we designed our campaign for this platform. We then took materials from the social media campaign to an international policy event, to pursue our new, wider goals.

How can you make your content actionable, shareable and rewarding? I'll take each of these in turn:

- **Actionable:** For gardeners, the action was simple — read our blog and sign a pledge to go peat-free in your garden. If we could get enough pledges, we would then use this to take our message to the mass media. Sadly, that never happened. However, we generated enough attention that I was contacted by an environmental charity working on a government taskforce to try and work out how to phase out peat-free composts from gardening and horticulture. At the time, industry representatives outnumbered other members 2:1 and an argument was made for me to join the taskforce as the only academic member. This happened as a direct result of the social media campaign and gave me a unique opportunity to feed evidence into the work of the taskforce, which is making strong progress towards phasing out peat-based compost in the UK. The second impact that arose from this campaign was more indirect, via the use of the materials we developed for the campaign at an international policy conference. This led to an IUCN resolution and to the IUCN team I was working with being invited to join a UN initiative to protect peatlands internationally. As a result of this, I am now able to feed research findings into this international group as I continue to pursue my global impact goals. Whether for gardeners, national policy-makers or the international community, we had a clear, actionable message: stop using peat-based composts.

- **Shareable:** To increase the shareability of our campaign, we worked on a number of evidence-based messages that presented our research in unexpected ways. We commissioned traditional infographics to support these messages, but during testing on social media these didn't get much of a response. We therefore redesigned the messages to appear on top of striking images of peat bogs, and immediately received significantly more online engagement (Figure 14). To craft our message as powerfully as possible, we used the marketing approach of making our message personal (rather than talking about peatlands, we talked about peat composts, which people were more likely to be familiar with and have a direct relationship with); unexpected (most of our audience were previously unaware that the products they were buying caused destruction to peatlands); visual (the images were chosen from stock photography websites to be visually striking, with strong colours and appealing aesthetics) and visceral (the facts we chose to highlight were quite shocking) (Box 12). We also created a campaign hashtag (#peatfree) and contacted a number of accounts with large followings to ask if they would consider adding the hashtag to their profile photo. We chose this hashtag because it was already in use by the sustainable gardening community, ensuring that people would find useful material via the hashtag even if we weren't providing new content.

- **Rewarding:** The main reward was information, but not only about the damage caused by peat composts. We provided information about how to find peat-free composts, and how to make sure they would perform well in the garden. Similarly, access to new, evidence-based ideas was the reward for members of the policy community who engaged with our team.

Figure 14: Messages from Project Maya's 2014 #peatfree campaign

Who can you work with to make your use of social media more efficient and effective? Project Maya identified a number of minor celebrities and charities who they thought might support the campaign, and reached out to each one individually, asking if they would put their name and image or logo on the website, endorsing the campaign. Each of these celebrities and organisations then linked to the campaign from their accounts, significantly boosting the reach of the campaign. The national and international policy impacts would not have happened were it not for the influence and reach of two major environmental charities who picked up on the campaign. The first was a charity that was approached to endorse the campaign, and the second was a charity who I had already targeted and begun working with as one of the most influential organisations working in the policy sphere around the issues I was researching.

Box 12: Make it PUVV

Successful online engagement is:

- **Personal:** Create designs with a personal hook in mind and ensure the campaign cultivates the feeling of personal relevance.
- **Unexpected:** People like consuming and then sharing new information. Work to pique their curiosity and reframe the familiar.
- **Visual:** It is important to show, don't tell. Use photos and visuals.
- **Visceral:** A campaign that triggers the senses and taps into emotions is much more likely to be successfully shared.

In conclusion, I hope this chapter has opened your eyes to the huge potential of the online realm to generate impacts from research. If this is something you are interested in pursuing further, I hope that the chapter has also made you aware of the unique risks you will face as a researcher in this realm. I have tried to provide methods that can enable you to mitigate as many of those risks as possible. For some of you, this chapter will have confirmed everything you previously feared about social media and reinforced your decision not to engage. My hope is that you continue this stance with pride, being able to justify your position to others more effectively than before. Whether you protect your time by disengaging from social media or by engaging with some of the techniques I've suggested in this chapter, my final plea is that you keep asking yourself one question, again and again: could I be doing something else more useful right now?

Chapter 18

How to generate research impact from Twitter and LinkedIn

In this chapter, I want to explain in more detail how you can use two social media platforms to generate impact from your research. I have chosen Twitter and LinkedIn because these tend to be used most widely to best effect by researchers from across the disciplines.

However, there are many other platforms available, and some are particularly useful for certain purposes, for example, Facebook for public engagement events, finding research subjects or engaging with otherwise hard-to-reach groups. If you are trying to reach younger people, you may want to target Snapchat or some of the other newer social networks. If your work is highly visual, then Pinterest or Instagram might work well for you. Ask yourself the first two questions from my social media strategy in the previous chapter and make sure you know what you're trying to achieve with whom, and select your platform accordingly.

How can Twitter enhance the impact of your research?

Twitter is one of the most powerful social media platforms for academics, given the number of highly focused and influential networks of people who use it. Effective use of Twitter doesn't just amplify your research, it allows conversations to take place about it. This can enrich your research and enable you to make a far greater impact.

1. Tweet yourself, your projects and your institution

In addition to your personal Twitter profile, consider opening accounts for some of your research groups or projects. Each of your research projects is likely to have a different focus, and you're probably a member of more than one group or institution in your university that doesn't have a Twitter account. A project Twitter account is an easy addition to your next pathways to impact statement when you're applying for funding, and some sort of engagement with social media is increasingly expected by reviewers. However, don't just add it for the sake of it — make sure that you have identified publics or stakeholders who are likely to preferentially engage with you on social media, and have clear impact goals you will pursue via a social media strategy.

Opening an institutional account will usually need to be a group decision. If everyone agrees, others can either send you material to tweet or you can give everyone the Twitter username and password to tweet themselves (if so, you'll need to agree on the nature of material you want posted, or it may be easier to decide on the things you want to avoid).

Open accounts for major research projects that will be going for a few years, and that you hope will have some form of successor project in the future (so you've got time to build a following and don't have too many accounts to manage). Again, the burden doesn't have to be entirely yours — it can be delegated to a post-doc and shared with other team members. Other ideas you might want to consider:

1. homepage, include the link in newsletters and presentations, and consider putting it in your email signature.
2. Every time you do a conference/workshop/seminar presentation, put your slides online (e.g. using SlideShare) and tweet them.
3. Every time you get a paper published, tweet the link to the article on the publisher's website (if it's not open access, consider adding that you can send copies if need be). If you can get permission, upload a copy on ResearchGate or similar and tweet the link.
4. Tweet quotes from speakers at conferences you attend, using the conference hashtag (make one up if there isn't one), to connect with other delegates and make them aware of your work.
5. Set up alerts (e.g. from Google News and Google Scholar) for key words and authors that are particularly relevant to your work, so you can be the first to let your followers know about new developments linked to your shared interests.
6. When you've got a tweet that is of much wider, general interest, you can retweet it from your other project/institutional accounts

to reach a much larger audience than you could ever command from your personal or one project account.

7. Next time you're revising your website, why not consider adding buttons to enable readers to share what they're reading via Twitter and other social media platforms?

2. Don't just wait for people to find you: actively promote your Twitter stream

There are some easy things you can do to promote your Twitter stream, like including links on your homepage, project websites and in your email signature. But more active promotion of your Twitter feed can attract many more followers:

- Make sure you've got an effective biography and enough really informative/useful tweets in your stream (typically with a link to more information) before actively marketing what you're doing.
- Contact relevant people with large followings to ask if they would retweet key messages you've sent — tweet or direct message them via Twitter, and if that doesn't work, find their email address via an internet search and email (or phone) them.
- Use popular hashtags (#) to make your tweets visible to more people (e.g. #PhDchat and #ECRchat). Notice which hashtags people you're following are using, and use them. If you're planning a Twitter campaign on a particular topic (e.g. linked to a new paper or policy brief), you could make up your own hashtag, but for it to work, others will need to use it, so you may want to work on getting a key tweet including your hashtag retweeted by others with larger followings.
- Have a growth strategy...

There is one growth strategy that is used by almost every organisation on Twitter that has an impact goal, whether that goal is profit or social good. Despite the technique making it into the peer-reviewed literature in 2016 (Schnitzler et al., 2016, *International Journal of Nursing Studies* 59: 15-26), most researchers have never heard of it. This isn't for everyone; most researchers do not need to become influential online to achieve their goals. However, if you have identified that social media is a potentially powerful pathway to impact with particular publics or stakeholders, you need to become influential. In social media land, influence = numbers.

So how do you do it?

1. **Have a social media strategy**: know what impacts you want to achieve through Twitter with which groups and come up with some indicators that will tell you if Twitter is actually helping you generate these offline impacts (see previous chapter).
8. **Set up a professional (project or thematic) account** from which you can promote research to specific audiences (and which you will feel comfortable promoting explicitly).
9. **Be credible and visual**: link to content and use images.
10. **Curate your top 3 tweets**: whenever you are leaving the platform for a while, make sure that your last three tweets (including a pinned tweet if you have one) effectively represent the best of what you put out from that account. To do this, look

to see which of your recent tweets got most engagement and retweet these to the top of your timeline.

11. **Only tweet when you've got something worth saying** (even if that isn't often): as a researcher, you are more likely to build a following and reputation if your content is of consistently high quality.

12. **Get the attention of influencers:** in your tweet, tag relevant accounts that have significant followings, send the tweet via a direct message to them, email them or pick up the telephone.

13. **Put your high-quality material in front of people who are looking for content like yours:** find others on Twitter who are generating similar content to you, and follow their followers regularly. You can assume that people who have recently followed a very similar account to yours are looking for high-quality material on the subjects you write about. Assuming your content is good, a high proportion of these people will follow you back once you have drawn your account to their attention. Many of them will retweet the content that made them follow you and many of their followers will like what they see and follow you too. Twitter may prompt you to confirm your password the first time you start using this strategy, but as long as you are generating good content and people are following you, Twitter will allow you to continue using this strategy because you are demonstrably adding value to the network and not a spammer. Depending on how well this works, you may hit a 'follow limit', but there are many websites and apps that can help you quickly unfollow accounts that did not follow you back, so you can continue using the strategy. As you follow increasingly more people, you will need to start reading your timeline from another account or from Twitter lists.

14. **Analyse your performance:** Twitter has built in analytics that will tell you which tweets are most successful — learn from what works and improve your practice.

3. Work on your signal-to-noise ratio

As a researcher, you need to build your reputation in your chosen field. Twitter can help you reach a network of highly relevant researchers, as well as potential users of your research, and make them aware of your work. To do this effectively, you need to decide what it is that you want to be 'known' for, and then work on building your reputation in that area. Most people will follow you because they share your core interests (your 'signal'), but they will rapidly lose interest if too many of your tweets are not relevant to these interests (effectively 'noise' they have to filter out when scanning through their timeline):

- Consider how useful and relevant each tweet is before sending it, to increase the likelihood that your followers find your tweets useful and keep following you.
- Ensure the majority of your tweets have hyperlinks to further information.
- Provide an image (or video) to accompany your tweets where possible (research by Twitter shows that tweets with images are retweeted 35% more than text-only tweets, and videos give a 28% uplift). Bear in mind that some web links automatically generate an accompanying image (e.g. many blogs, newspaper sites and video sites automatically generate an image, title and first line of the article below your tweet once it has been sent).
- Avoid sending too many tweets and retweets at a time — if you're at a conference and tweeting every couple of minutes, followers who aren't interested in the conference are likely to get fed up with you dominating their timeline on a single narrow issue and unfollow you.
- Avoid using too many acronyms and abbreviations in your tweets — they may make sense to you but many people reading fast will simply skim over your tweet if they don't understand you instantly. It is better to say less in complete words than to try and cram too much in if it means you resort to acronyms and abbreviations.
- If you're increasingly tweeting about things that are very different from your core interests, consider setting up a new Twitter stream devoted to that issue/interest.
- If you're tweeting from a project or institutional account, try not to mix work and personal tweets. Remember you're tweeting on behalf of a group, so telling people about what you're doing on holiday is going to sound a bit strange (either your institution appears to be on holiday or it becomes clear that the Twitter stream is really only about one person —who's on holiday — and not the whole group). If you do want to mix personal and work tweets (some commentators suggest this can help build rapport with your followers), make sure your biography clearly states the name of the person tweeting on behalf of the project or organisation.

- If you find that you've started automatically skimming or skipping tweets by certain people, the chances are they rarely have anything particularly relevant/useful to say — mute or unfollow them and reduce the amount of noise you have to deal with.

4. Get your timing right

Because of the way Twitter works, most people only read a fraction of the tweets in their timeline, so if you're tweeting on a day and time that none of your audience are reading their timelines, you could be tweeting into the void (for example, tweeting in the night might be useful if your primary audience is on the other side of the world, but if not, then your tweets will probably be lost deep in your main audience's timelines by the time they wake up and start reading tweets over breakfast). Timing is also about linking to the issues of the day — reframing to link into an ongoing news story or debate can really get your research some attention:

- Link your tweets to ongoing events in your discipline and the news, using linked hashtags where relevant
- If you've got a lot to say, don't tweet in bursts; rather spread your tweets through the day, using something like HootSuite to automatically schedule your tweets to be sent at different times of the day and week (so you don't have to keep interrupting your day). Someone who only logs onto Twitter at the end of the day may not get to the three tweets you put out at 8 a.m., but will probably get at least one of the ones that were scheduled for the afternoon. Warning: your friends might think you have super-human powers when they discover you're tweeting while lecturing or speaking at a conference (many people actually schedule conference tweets in advance, based on the programme timings)
- Get to know when your followers are most likely to read your tweets and put your material out at these times. For example, from experience I have found that @fasttrackimpact followers are most likely to engage with material between 8 and 9 a.m. on weekdays, so this is when most of my tweets are posted
- To increase the chances of people following you when they look through your tweets, I like to avoid the repetition of sending copies of the same tweet at different times of day. Instead, I will retweet key messages at different times of the day, and on successive mornings, to make sure they appear in the timelines of most of my followers
- Although this goes against all other advice for Twitter users, I think that quality is more important than quantity for researchers who want to build a strong reputation online. Rather than trying to tweet at least once a day, as many people recommend, I tweet when I have got something useful to say. As a result, I may send less than ten tweets in a month, retweeting my most popular previous tweets, and making sure I leave my best three tweets at the top of my timeline as a 'shop window' for anyone who comes across my account when I am leaving it for a few days or weeks.

5. Use Twitter as part of a wider social media strategy and impact plan

Twitter is just one of many social media platforms, so consider putting your material out via other platforms too, and remember that people who might use your research aren't always using social media, so you're going to want to think about other ways of reaching out to your audiences:

- Come up with a properly thought-through social media strategy as part of a wider impact plan for your research, whether as an individual, a project or an institution (see previous chapter).
- Adapt your approach to each platform, e.g. I will see how my messages resonate on Twitter, and only put those that get significant engagement out via the Fast Track Impact Facebook page or LinkedIn. I will sometimes add a personal twist to my own Facebook posts about my work, which are few and far between. On LinkedIn, I will sometimes add more detail, for example, linking to the paper, blog and video on separate lines.
- Remember that social media is just one form of communication, and that there will be many who are interested in your work who are not using these technologies. Keep up your newsletter — printing and posting where relevant (but still tweeting the link to the PDF, hosted somewhere you can count hits like Scribd, ResearchGate or Issuu). Keep presenting at conferences and running workshops for the end users of your research (tweeting videos of what you do on YouTube and putting your presentations on SlideShare, of course).

6. Constantly refine your practice

Watch how other academics, projects or institutions with large followings tweet:

- Learn good practice from others, and experiment yourself.
- Take note when something annoys you about the way other people use Twitter and avoid doing that yourself.

Monitor and learn from your successes and flops:

- Which of your tweets are most likely to get retweeted? Which tweets don't get retweeted? What do they have in common, and what can you learn from this? How were you using Twitter on the day you got 10 new followers?
- Put (open access) documents that you cite on Twitter in places where you can count hits — which tweets make people click on

the link (and presumably read your document), and which ones fall flat? What can you learn from this?

- Experiment with different headlines in Twitter to see which ones work best — try and reframe your point and tweet it again later that day, and see if you have more success.
- Read through the material you're tweeting and find quotes you can use to promote the link in a slightly different way — sometimes one of these quotes really takes off, far more effectively than the headline. If you're tweeting a blog you wrote, then you might want to consider retitling the blog at this point!

7. Remember it's all about relationships

Don't forget that Twitter is about communicating and building a relationship with people and not just marketing your own or your institution's work at them. So, remember to check other similar institutions'/academics' tweets and respond to those that are interesting. Twitter allows your work to reach a much wider audience and also enables more discussion of your work with others who may put it into practice.

Also, as with any other social setting, there is 'Twitter etiquette', for example, if someone gave you the information you are tweeting, credit him or her with it, either by using "via @person1" (if they are a Twitter user) or as a quote next to their original tweet.

Using LinkedIn for research impact

LinkedIn has unique capabilities you can harness for achieving research impact. It is particularly effective for engaging with stakeholders, rather than publics, because of its focus on professionals. If you have done a publics/stakeholder analysis and know which groups will benefit most or have most power to facilitate your impact, LinkedIn has a powerful search function that will enable you to target your message to key stakeholders based on city or country, the organisation they work for (or previously worked at), the sector or industry they work in more broadly, or their interests.

I will explain how I run social media campaigns via LinkedIn using an example from my own research to illustrate. Like any good social media campaign, you need to start with a clear impact goal and know your audience. In my example, the impact goal was to get private investment in peatland restoration. Our market research had identified a number of business types that were more likely to be interested in the opportunity than others, and so we used LinkedIn to target anyone with a job title including 'corporate (social) responsibility' and 'sustainability' in large companies from the relevant sectors:

- The first step is to connect with stakeholders from organisations you would like to work with to deliver impact. Use the features in the advanced search on LinkedIn — it is a surprisingly powerful search engine. You don't have to have worked with them before to be able to connect with them (though in some cases you will need to Google their email address to make the connection request). Use your current job title in your connection request, and include a short note (rather than just using the default message), explaining that you're interested in their work and think they might be interested in your research (you'll get a much better acceptance rate this way). When I did this for my peatland research, over 90% accepted my connection request.

- Once you've got connections with a good range of stakeholders, create status updates and blogs on LinkedIn specifically related to the impacts you want to achieve. In this way, you are starting to put your work in front of people so they become more familiar with it and are more likely to trust you as a credible source of information and help. LinkedIn Pulse blogs are highly visible on the platform so consider using this. If you are blogging elsewhere, consider copying the first paragraph and image to a LinkedIn Pulse blog, and then linking from there to the original post. In my case, I linked to previous blogs in this way and wrote a new blog linking to an event that was coming up (that I would later be inviting people to via LinkedIn).
- Find LinkedIn groups talking about issues linked to your research, request to join and contribute to the discussion before adding in links to your work.
- After some time of generating content in this way, and making yourself visible to your new contacts on the network, start interacting with your LinkedIn contacts by sending them messages about your work. These go straight to their email inbox and they can reply from their email, so it's easy for people to respond to you. According to your social media strategy and impact plan, you should have a clear goal in mind when you are reaching out to people. Ask yourself what you want to achieve and what they will gain from interacting with you. In my case, I was inviting them to an event where they would be able to find out more about my work, as well as a lot of other relevant stuff (it was part of a wider conference). I was pitching a specific side event, where we would be putting forward our investment opportunity. Interestingly, I was given the opportunity to sign up for a 30-day free trial of LinkedIn's premium product at this time, and used this to send InMail messages to around 50 additional contacts who were not connections of mine. Not a single one replied. The secret of the (free) approach I took was that I had subtly built trust with my audience before approaching them with my request.

Our social media strategy worked, and we got lots of relevant people at our launch event where we pitched our investment opportunity. However, we failed to get impact from this event as these people informed us that the decisions we were asking them to make could only be made by Chief Executive Officers. At this point, we switched strategy to an offline alternative pathway to impact to get to this hard-to-reach group. At the time of writing, we now have over 20 projects funded across the UK.

Finally, if you want to see how I use Twitter and LinkedIn to drive impact, or simply want to discuss anything you've read in this book, you can find me on Twitter @profmarkreed and @fasttrackimpact or search for my name and Newcastle University to find me on LinkedIn.

Chapter 19
Presenting with impact

A PhD student stands nervously on stage as 250 students noisily pack into the lecture theatre. It is his first ever lecture and he stayed up half the night preparing it. Through the fog of tiredness he can't quite remember how he planned to start the lecture, but he makes a stumbling start nonetheless. Soon, however, it becomes clear that the things that made sense as he was putting the slides together during the night no longer make sense now. He grinds to a halt at a slide with a graph he can't interpret. The pause turns into a silence, and the silence seems to go on for an eternity. He presses the arrow key. Next slide. He doesn't even remember putting the next slide into his presentation. Panic begins to rise as he desperately searches his memory for what this part of the lecture was meant to be about. He presses the arrow key again, 250 pairs of eyes silently searing into him. By now his heart is pounding wildly and he is sweating profusely. Next slide. Next slide. The words on the slides have turned into meaningless symbols. And then finally, the silence is broken — but not by him. One of the students is heckling him. He makes the mistake of trying to explain himself, and a student in the front row gets up and walks out of the lecture theatre. He follows shortly after. Lecture over.

Fast-forward five years. The same PhD student, now a lecturer, is in South America and is running late for a presentation he is giving. This is a very different presentation. Instead of students, he is talking to representatives of every country in the world at the Conference of the Parties to a UN Convention. He can see the hotel on the other side of the train tracks, but every street he runs down is a dead end with no way across the tracks. Panic starts to rise again — a train track is between him and the opportunity of his career. Eventually, he finds a way across and gets to the conference venue with less than five minutes to spare before he is meant to be speaking. But there's a problem. To get into the inner chamber where he is speaking, he needs special security clearance, and the badge he has been given won't give him access to that part of the conference. He is so out of breath that he can barely speak and the security guard doesn't speak good English. Eventually, he manages to speak to the head of security, and points to the name on his badge and his name on the programme. Then he points to the time printed next to his presentation, and to the clock on the wall, which is now showing the time his presentation was meant to start. As he enters the room, the Chair is already introducing him, and by the time he reaches the podium, it looks as though it has been perfectly choreographed. He has even got his breath back. He looks into the cavernous hall at the rows of desks with people sitting behind country nameplates, takes a deep breath, and gives the presentation of his life.

Power to impact or alienate

I've told you these two stories about me, because I believe that no matter how terrified and unconfident you may feel, it is possible with a few tips and some practice, to present your research with real impact. Personally, I think that there is only one thing that separates the 'superstars' of our disciplines from the rest of us. It is not that they are any more intelligent than us, or doing better work. It is simply that these researchers are better communicators. They have worked out how to take the same data we produce and turn it into a highly cited paper and a plenary talk that inspires audiences around the world.

The same applies when it comes to non-academic impacts. Some researchers make talking to businesses and policy-makers look easy. The rest of us look on in awe, desperately wondering how we could create such succinct and relevant messages based on our research. Many of us conclude that it is "easy for people who do that sort of research". However, most of these people started in a similar position to us; they just focused for a while on an aspect of their research that had the potential to be useful to that audience, and spent some time thinking about how they could communicate it powerfully.

I regularly hear people from business, the third sector and government complaining about researchers' love of complex graphs, which they suspect are designed to make the researcher look clever and make them feel stupid, because they weren't ever put on the screen for long enough or made large enough for anyone in the audience to actually read. Presentations like that don't just squander opportunities to build relationships and generate impact; they alienate the very people who might be able to benefit most from your work.

Researchers have to do public speaking on a regular basis, whether it is to other researchers, publics or stakeholders. The crazy thing, however, is that most of us are never given any proper training. I have been lucky enough to receive training from a professional voice coach who works with politicians and other public figures, and I have picked up lessons from colleagues along the way. Out of everything I've learned, I believe that there are five key things you can do that will transform the impact of your public speaking. Given how much I've needed this help, I thought I'd share what I've learned with you, so that you don't just get your message across, you transform and mobilise your audience.

1. Have purpose

The first minute of your talk is make or break time. Based on what you say in your first minute, your audience could either be glued to your every word, or pretty much dismiss everything you say in your whole talk. To engage your audience, there are just three things you need to do in your first minute:

a) **Establish your purpose and the benefits your audience will get from listening to you:** most of us know that we need to start a talk with our aims. I'm suggesting you should just have one single purpose that people can instantly understand and remember, and very quickly explain the tangible benefits that your audience will get as a result of achieving this purpose (even if those benefits are just learning something new). Finally, put yourself in the shoes of your audience and ask yourself why your purpose, and the benefits you've identified, are likely to be important to them. Then actually explain the benefits of listening to your talk to your audience.

b) **Explain who you are and why your audience should listen to you:** you don't have to be the world expert on your topic, but there must be some reason why you are talking and not some random stranger picked off the street. What sets you apart from that random stranger? What credentials do you have? Why are you passionate about this topic? There is a fine line between establishing credibility and boasting, and you need to be careful not to alienate your audience by giving them your CV. However, there is good evidence to show that audiences are more likely to listen to and learn from speakers that they deem credible, so it is important to establish this in the first minute of your talk.

c) **Signpost what is coming next:** people like to know where they stand. You shouldn't spend much more than a sentence doing this (don't spend half of your talk going through your plan and explaining what you're going to do). Just explain the key sections or steps you will go through to reach your purpose, so your audience feels able to relax into what is about to happen.

2. Connect

The best speakers empathise with their audiences, and their audiences identify with them. Opening a channel of empathy with a stranger can be a huge challenge; doing this with a room full of people you don't know is much harder. However, there are four quite straightforward things you can do to establish empathy with any audience:

a) **Know your audience:** do your research so you know who is going to be in the audience and why they have come. Be aware that there may be quite different segments to your audience, who are looking for different things from you. If you are not able to research your audience, then take some time before you speak to sit next to someone in the audience and find out why they are here and what they are hoping to get out of the event. You will have to assume that their answers are broadly representative of the rest of your audience, but at least you are not going in blind. Once you know something about your audience, you can adapt what you say in your opening minute to make sure you are connecting with benefits that this particular audience is likely to value, or you've explained the benefits in a way that makes it clear why these should be important for this particular audience.

b) **Use powerful stories:** we all know the power of stories to convey complex concepts in memorable ways, but not all stories have equal power. First, think of a few stories that are relevant to the one single purpose you identified in your first minute. They may be directly relevant or they may be a metaphor that you feel sums up your purpose powerfully. Then identify the story that addresses as many of the four points in this list as possible: **Personal** stories help open a channel of empathy, showing that despite being up on stage you are only a person, with weaknesses and passions, just like them. Stories that demonstrate some degree of vulnerability show that you trust the listener, and they are then more likely to warm to you and trust you themselves. If you can, try and include something **unexpected** in your story, to catch your audience's attention, help them remember your story and make it more likely that they subsequently share the story with others. If you can paint a **visual** picture with your story, whether in the mind's eye or through images, your audience is more likely to be able to recall your story, and if the image effectively illustrates your story, it will add real impact to what you are saying. Finally, engage to some extent with your audience's **feelings**. This doesn't need to be anything particularly dramatic, but stories that rouse some sort of emotion are more likely to stick than stories that leave your audience cold. If your story is strongly linked to the core purpose of your talk, then by remembering your story, your audience will remember your purpose, and from there, much of the content of your talk. As an example of a story that I think ticks these four boxes and is linked to the core purpose of this article, look at the two linked stories I told at the start about presentations I gave to students and to the UN.

c) **Ask 'you-focused' questions:** asking your audience directly to put themselves in your shoes can be a powerful way of establishing a channel of empathy with them. This may be difficult for many research-based talks, but with a bit of imagination, it may be possible. For example, "What would you do if..." or "What would you think if I told you...".

d) **Use empathetic body language:** it is possible to become a more empathetic speaker simply by making your body language more open and approachable. Consider choosing clothes that do not emphasise any differences between you and your audience (for example, I often remove my suit jacket when training PhD students), avoid closing your body language, and adopt a positive and energised posture that shows your audience that you are putting in effort and really value them. You will often discover that your audience starts to mirror the emotions you are projecting through your body language, and will start to feel more open, trusting, interested, and energised by your talk.

3. Be authoritative and passionate

A lot of people avoid looking authoritative for fear of looking intimidating, but these are two very different things. Someone who is genuinely authoritative will typically embody a quiet confidence that does not need to boast or intimidate. Similarly, many people avoid being too passionate for fear of sounding like a salesperson or politician. Someone who is genuinely passionate about their subject, however, will typically exude their passion without even trying and their audience will find their enthusiasm infectious.

There are three very simple things any speaker can do to demonstrate authority and passion:

a) **Be aware of your feet:** look at yourself consciously next time you give a talk, and see what your feet are doing. Some people pace; others step backwards and forward as they speak. Some people sway; others do a bit of a dance as they speak. All of them do it subconsciously and, without realising it, have a subconscious impact on their audience. As we move around, we are likely to distract our audience from what we're saying, look less confident and create a sense that our words are insubstantial. On the other hand, speakers who have their feet firmly on the ground in one place are perceived to be focused, confident and substantial. This doesn't mean you have to stand like a statue, but you need to use movement strategically. Choose a 'home' position from where you can introduce your talk and your core purpose (usually this is somewhere fairly central). Then have a number of 'stations' around the stage (for example, to the left and right of your screen) where you can move between points, to keep your audience's interest and make clearer distinctions between points. Then, at the end, return to your 'home' position to make your conclusions and fulfil the purpose you set out to achieve.

b) **Be aware of your hands:** what you do with your hands can be similarly distracting and undermining if you are not aware of them. Putting your hands in your pockets may suggest a level of informality that makes it look like you're not serious. Clasping them behind your back may make you look suspicious, like you're hiding something. So what do you do with your hands? Simply clasping them in front of you is a safe bet if you're nervous, but you will probably look nervous as a result. Using lots of flamboyant hand gestures may be very distracting for your audience. My voice coach therefore told me to draw a TV shaped rectangle in front of me, and to keep all my hand gestures within that rectangle. She also advised against any kind of aggressive gesture, such as pointing, preferring a small number of open and inviting hand gestures. Now, my hands aren't ever going down to my sides and drawing people's attention away from my face; all of my gestures bring people's eyes back to my face and my message. By using confident but muted gestures, I look credible, in control and confident, and can use my hands to add emphasis to my points and convey my passion.

c) **Use emphasis to make every word and sentence count:** if you're going to say something, make it count. Make every single word count. If you trail off at the end of a sentence because those words aren't actually important, don't say those words. Cut out the unnecessary words and then speak every single word in that sentence with equal conviction. If you find yourself skipping over a sentence fast or mumbling some context or explanation, ask yourself why you are saying those sentences. If they are in fact important, then don't skip over them in the way you speak them, or your audience will skip over them in their attention, and ignore your point because you effectively told them it wasn't important. Again, if it isn't important, just cut it. Now, once you've learned to make every single word of every single sentence in your talk really count, consider how to put emphasis on the key point of each and every sentence, to demonstrate to your audience why it matters. You may want to use pace, slowing down and spelling out key points, or pausing before or after a key point, allowing it to sink in. You could use volume (sparingly) or vary your tone of voice more than you naturally would in conversation. Many researchers object at this point because it all starts to feel a bit fake. The last thing we want is for our audience to think we are insincere. However, most audiences expect people to speak slightly differently when they are on stage than they do in conversation, just as your family expects you to speak differently to them than you do to your colleagues at work. Your audience is far more likely to appreciate your more interesting and engaging style than it is to complain that you didn't sound exactly like that when they spoke to you in the break.

4. Keep it simple

The most common mistake that researchers make when presenting is to make their talk too complicated. Most of us can be forgiven for falling into this trap because our research is usually by definition fairly complex. However, the most successful communicators have spent time thinking about how they can communicate their complex research in a way that is deceptively simple, and they will do so around a single key message, which they make as memorable as possible:

a) **Find a single memorable message** linked to the core purpose you identified in the first minute (in some cases it will be the same thing).

b) **Present your key message early and revisit it from many different angles:** if it is not presented during the first minute, then it should be presented in the first section of your talk. Then revisit it from different angles throughout your presentation, using metaphors, stories and images where you can, to make your point stick in people's memories.

c) **Link all your subsequent points back to your key message:** having a single key message doesn't mean you can only speak about one thing in your talk. However, it is important to remember that most people will only remember a fraction of your talk, and you have put in effort already to make sure that they remember the most important point. If you then clearly link each of your subsequent points back to your key message, then your audience is much more likely to remember these other points when they recall the key point. Rather than having to remember many different stories and plot lines, they only have to remember a single story and plot line that logically flows from the memorable story or image you used to introduce your talk.

5. Polish your talk

Finally, make sure you don't undermine the credibility of your talk by stumbling over your words or doing your design en route to the venue. Make your talk look and sound polished and your audience is much more likely to think you are credible and trustworthy, and listen to what you have to say:

a) **Practise and practise again**: the most polished presenters make public speaking look easy. I think some people think that one day they will automatically be so filled with confidence and expertise that they will just be able to stand in front of an audience and speak like a pro. The reality is that those who make speaking look effortless are usually those who have put the most effort into it. Practise in front of the mirror. Record yourself on your smart phone. Practise in front of low-stakes audiences and seek their honest feedback.

b) **Watch out for verbal fillers**: many of us are plagued by unconscious verbal fillers that we are totally unaware of, for example, ums and errs, "you know", "like" and "sort of". These verbal fillers can have a couple of unfortunate consequences. First, once your audience notices them, your verbal fillers will start to really annoy and increasingly grate on people, distracting them from your message. Second, speakers who use more verbal fillers are more likely to be perceived as nervous by their audience. It therefore follows that even if you are in fact very nervous, you can instantly create an impression of calm confidence by replacing your verbal fillers with silence when you need thinking time. The use of "you know" is a common and unfortunate verbal filler among researchers who are often talking about things that most people don't actually know about. They therefore implicitly make their audience feel inadequate, as though they really should know about the things they are being told about.

c) **Use your visual aids to add impact, not as your notes:** how often have you been to a high-impact plenary talk that is full of dense text on the screen? If you've not ever been to a particularly high-impact talk, take a look at researchers doing TED Talks at www.ted.com. If you can't think of a single image to support your talk, then don't use PowerPoint at all. The problem with the way that most researchers use PowerPoint is that the audience is too busy reading the text on the slides to properly hear what the presenter is saying, and they can't properly take in what they are reading, because the presenter is saying something slightly different and out of synch with what they are reading.

The art of presenting is under-taught and under-valued in academia. However, by learning and practising a few simple techniques, you will be surprised how much more effective you can be. Creating a talk that truly inspires change will take time, but many of your greatest opportunities to achieve impact from research may arise from the power of your talks, and the distance they start to travel.

Chapter 20
How to engage policy-makers with research:
a relational approach

One of the greatest privileges of being a researcher is the potential we have to operate in a sphere of influence far beyond many other professions. A water engineer might provide drinking water to thousands of villages during their career, but by working with governments we have the potential to influence policies that can bring water to many millions more people. In this chapter, I want to discuss how you can work more effectively with the policy community to generate impacts from your research.

By 'policy community' I don't just mean policy-makers. I am talking about the diverse network of people who feed into the development and implementation of policy, including politicians of all levels (from backbenchers to ministers, both members of the government and opposition parties), civil servants (including those working in both evidence and policy roles in government departments, agencies and other governmental organisations), and the dynamic group of individuals and organisations that shape policy as they move in and out of spheres of influence at different points in the policy process (including, for example, third sector organisations, unions, consultants and lobbyists).

The aim of this chapter isn't to give you a detailed guide to the political apparatus of any particular country — for that you will need to look elsewhere. My goal, instead, is to get you to think differently about how you engage with the policy community, and to persuade you to consider taking a more relational approach to your work in this sphere. In this way, I believe we can become more influential, and increase the chances of our work generating impact. I am not suggesting that we abandon the traditional ways of engaging with policy. Instead, I am suggesting that we don't stop engaging once we've submitted our consultation response or given evidence to a committee. For most of us, what happens next to our evidence is a black box. We may eventually see our work cited in a policy document that leads to impact, or we may never hear anything further. I am suggesting that we do what we can to enter into that black box and help colleagues in the policy community work with our evidence to address the challenges they are facing as they develop policy. This approach may create risks to our time and reputation if things don't go according to plan. In this chapter, I want to make you aware of these risks, as well as the opportunities of taking a more relational approach, and show you how you can mitigate some of these risks to have greater influence on policy.

There is no such thing as evidence-based policy

I'd like to start with a bold statement. There is no such thing as evidence-based policy. For this to exist, policy-makers would need to base policy on evidence. However, evidence is often highly fractured, providing evidence about a single part of the problem in a specific context, or describing future environmental impacts based on natural science alone, without considering social, cultural or economic factors. Evidence may be uncertain, providing competing claims based on different methods at different scales. Few citizens or politicians have sufficient technical understanding of our research to be able to critically evaluate competing claims and counter-claims, making it easy for lobbyists to sow confusion by amplifying uncertainties.

As a result, members of the policy community must interpret often contradictory research findings, alongside other lines of argument put forward by people with competing ideologies. Policy-makers must therefore consider moral and ideological arguments alongside practicalities (such as budget constraints) and unpredictable external events that constantly change the parameters of the decision being made.

Some have described the relationship between research and policy in more cynical terms, as a way for governments to legitimise policies with reference to evidence from research only when it supports their politically-driven priorities. As J.M. Keynes put it, *"There is nothing a politician likes so little as to be well informed; it makes decision-making so complex and difficult"*.

It is easy to sit on the sidelines and criticise colleagues in the policy community for the many imperfections of real-world policy processes. It is a lot harder to be critical if you have spent any time working in government departments, trying to juggle the multiple competing claims on your time and the curve-balls that get thrown at you by politicians or external events. In addition to synthesising evidence from research, there is the need to balance the interests of different stakeholders and public opinion, and listen to the practitioners who may explain why theory (from our research) doesn't always translate into practice.

One response to this complexity is to defend the primacy of scientific knowledge as the only way of finding rational argument and universal truth upon which policy can be based. In response to the conflicting accounts often provided by science, we simply need more and better research.

An alternative response is to accept that pragmatic and ideological considerations will probably alter little in response to more and better research. Instead, we move from trying to achieve evidence-based policy to seeking evidence-informed policy. We become knowledge brokers, using the widest possible body of evidence to provide evidence-based options. While it might appear that the evidence is stacked in favour of one option, we empower a policy-maker to choose an alternative option in the full knowledge that the evidence suggests there will be trouble ahead. In a world of evidence-informed policy, our task is to ensure that the evidence is available and on the table, in forms that are just as palatable and persuasive as the arguments being proposed by others for options that we know from the evidence are likely to be fraught with difficulty. If we care about getting our evidence onto the decision-making table, we need to learn how to become more influential. We can't just submit our evidence and hope for the best.

Combining bottom-up with top-down approaches to influence policy

As a researcher, I want to make evidence accessible to policy-makers in an engaging and influential way. The word 'influence' in this context is problematic for many researchers, but if we want to take a relational approach to impact, I believe that it isn't enough to simply create a policy brief and put it online. The reality is that the majority of people in the policy community call on trusted advisors for advice relating to research evidence, and are less likely to listen to evidence from sources they do not trust. Just having your paper published in a top journal isn't enough to engender trust and be listened to. You need to demonstrate your credibility and trustworthiness in the context of a long-term relationship with key members of the policy community and become embedded in that community if you really want to be listened to.

I believe that one of the most effective ways of achieving policy change is through a 'pincer' movement of influence from the bottom up and the top down (Box 13). It is usually easiest to start from the bottom up, connecting with policy analysts and government researchers who have a similar background to you, and who are likely to easily understand the research and where you are coming from. Starting by building strong, trusting relationships with more junior civil servants, you can begin to understand which of their managers have relevant interests and influence, and begin to introduce them to your ideas too.

However, this approach can only go so far if the top decision-makers (e.g. ministers) are not aware of your work. Getting access to these top decision-makers is a rare opportunity for most researchers, so you may need to rely on intermediaries, such as charities or others, who have existing relationships and routes to those in power. I've discussed some of the ethical dilemmas that this poses for researchers below. If you can present a case for policy change based on your evidence (even if second-hand via an intermediary), and convince a senior policy-maker that they should take action, then it is important that they are met with informed civil servants when they take the idea to their team. If their team hasn't heard of your work, doesn't trust you and isn't convinced by the case as it is put to them by the minister (which may not be how you would have put it to them), they may raise so many questions and doubts that your ideas are dismissed as unworkable. On the other hand, if the minister is met with informed judgements from civil servants who are already aware of your work, and have critically examined it, there is a much higher chance that change will occur. Equally, just convincing civil servants that your research deserves attention may not be enough if it doesn't fit with the interests and priorities of the minister at that time. So taking both the top-down and bottom-up route is, I believe, important if you really want to effect change. Box 13 provides a few questions that can help you design your own 'pincer' movement for evidence-informed policy change.

Box 13: Designing a pincer movement for evidence-informed policy change

1. Identify policy stakeholders from your stakeholder analysis (see Chapter 13 and template in Part 4). Check that you are being as specific as possible: which policy area, department or team are you identifying that might be interested in your research?
2. Identify areas of policy that may be related to or similar to your research in some way:
   a) Can you link your research to these live policy debates in some way? Would the insights from your research enrich these debates?
   b) If so, who would you need to collaborate with to connect to these wider debates?
   c) If not, what future work might you do that could contribute to these debates? What could you do now to start this work?
3. Top-down influence:
   a) What other organisations are working in this policy area to influence policy?
   b) Which of these do you think has most influence?
   c) What are the key messages from your research that are likely to be of most interest to them?
   d) Can you find out more about their priorities and modes of operation, and start to get to know people in the organisation who will be interested in your work?

4. Bottom-up influence:
    a) Which evidence teams within the civil service are working on the policy debates you can connect to?
    b) Can you make a policy brief that is relevant enough to secure you a meeting with someone junior?
    c) If not, can you get introduced by someone who they already know and trust (look through your network and those of your colleagues and work out what you could do for the person who might introduce you).
    d) Once you have a contact within government, find out from them what evidence gaps they need to fill and offer to help.
    e) Stay in regular touch and build trust, asking questions that will enable you to work out who in their team and wider network has most influence. Find out about the events that these people go to and try to connect with them there so they know who you are and what you're doing before you are introduced to them by their own colleagues.
    f) Gradually connect your research with people of increasing influence via departmental seminars and one-to-one meetings.
5. Plan for your impacts: go back to your impact plan (Chapter 10, Table 3) and revise your activities and timings for engaging with policy stakeholders

How should I start?

The first step is to identify the key messages from your research that are likely to be relevant to current or future policy, and why these messages are important. This is often an iterative process, researching the policy environment and getting feedback from people in the policy community, to help you focus on the most relevant aspects of your research and frame clear messages that are likely to resonate with the issues and challenges they face. This initial feedback may be via social (or other) media or via people at the periphery of the policy community e.g. researchers who have a long track record of working with the policy community in your field, government researchers or agency staff. It is better to get constructive feedback from these people to have a polished, concise and relevant pitch ready for those who are likely to have greater influence.

If you are working on a fairly narrow topic (a common problem for PhD students who want to work with the policy community), it can be hard to make your work relevant enough to warrant attention from busy policy analysts. However, if you are able to make connections between your work and the work of colleagues, and contextualise this within the latest research findings that link your narrow research topic to the bigger picture, then it may become easier to reach these audiences. Although your research may now be reduced to a box or a paragraph and accompanying figure, at least there is a good chance that people will engage with it now.

Consider exactly what you might want a policy-maker to do with the knowledge you are providing — make sure it is something that is actually achievable, and if it isn't, then work out what the initial steps might be towards the action you'd want to see in the longer term. There is evidence that research findings that build on rather than break down existing policies are more likely to be adopted — recommendations for a series of incremental changes rather than a single-step change are more likely to be adopted. Having said this, sometimes it may be as important to enable an individual or organisation to 'unlearn' certain accepted concepts and ways of doing things (such as the accepted health effects of a particular food or lifestyle choice) in order to take on board new understanding based on the latest research (e.g. suggesting that what we previously thought was healthy may have negative consequences for health).

One of the greatest challenges of constructing messages from your research is how to communicate complexity and uncertainty clearly, without putting off policy-makers who want definitive answers. It is important to avoid giving a false sense of certainty e.g. via numbers, graphs or maps that hide variability, error bars or alternative scenarios. However, case studies, stories and personalised findings can help communicate complexity and bring the key points home to decision-makers. A common problem that members of the policy community have with researchers is our propensity to selectively promote our own latest research, overlooking equally valuable and often highly complementary work by other researchers that could significantly increase the value of our own research for policy-makers. By summarising other research on the topic, you may also be able to reduce uncertainty and increase the credibility of your own work by showing the range and depth of research that backs up your claims.

It can often pay dividends to work with professional communicators (e.g. science writers, knowledge brokers, your institute's public relations officers and/or film-makers) to translate your work into terms that can be understood by those you want to influence. Also, knowledge brokers can help facilitate your dialogue with policy-makers, helping you 'translate' discipline-specific language and mediate if necessary.

When should I engage?

The best time to engage is at the start of every research project. After you have identified your 'target audience', you need to find out if these groups really do find your research relevant to their work. Together, you will then be able to formulate research questions that are relevant for both of you. By doing this, everyone involved knows what outcomes are likely to arise from the research (and when), and potential uncertainties can already start to be communicated at this stage.

There are certain times when a piece of evidence may be crucial in policy decisions. It is therefore important to be in regular contact with members of the policy community, so that you can easily identify those key moments and changing demands. If they already know you, they're likely to come to you for the answers. Even if that means that you are being asked for evidence before the research has been completed, remember that you have a much broader knowledge base than the project you are currently working on, which could still enable you to link to existing published evidence to help provide the answers that are needed. In some cases, it may be possible to provide preliminary findings, as long as the limitations and uncertainties are made clear. New political leadership in a particular government department or agency can be a problem (in terms of continuity), or in fact may become an opportunity to present new ideas to leaders looking for new ways of achieving their goals. Working with political parties to get your ideas into election manifestos can be an effective way of getting research into policy, if you don't feel too uncomfortable about appearing to be affiliated with a particular party.

Where should I engage?

The majority of key players are extremely busy, and you need to consider how to bring your message to them. Most government departments and agencies will host seminars if you can demonstrate that your research is of great enough relevance, for example, by bringing together a number of key experts to present their research alongside yours. You can also hire a venue near the parliament and offer a free lunch to incentivise attendance. However, for many of the most important players, you may need to arrange a short face-to-face appointment with them or their close advisors to get your message across. Many politicians are active on social media, and this can be an easy way to get their initial attention and start to build relationships with them.

To engage with policy-makers on international policy matters, you will need to explore events and bodies relevant to your work, such as the UN Convention of Biological Diversity and its associated Subsidiary Body of Technical and Technological Advice. Remember, it is these technical events where many decisions are typically made and where you have greatest influence as a researcher, rather than the larger, better-publicised events which the high-profile politicians attend. Some countries have set up specific science-policy interfaces or platforms to enhance dialogue between researchers

and policy-makers. It is worth checking if one exists for your given research area.

Identifying who has the power to affect policy change

Using publics/stakeholder analysis (Chapter 14), it should be possible to identify organisations and key individuals within those organisations who are particularly influential, who you might want to try and build relationships with. These may be policy-makers themselves, or it may be the advisors who work closely with them within the civil service. It is important not to overlook organisations and individuals outside the policy community who have long-standing relationships with key members of the policy community and may have a lot of influence, for example, non-governmental organisations, charities, think tanks, business and lobby groups.

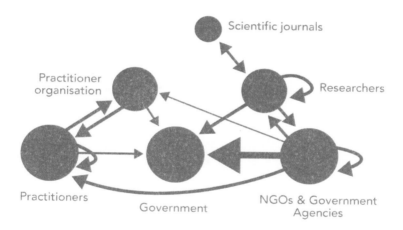

Figure 15: Simplified representation of a Social Network Analysis, showing how peatland research reaches policy-makers. Circles represent different sources or users of knowledge, with larger circles more likely to provide/receive knowledge than smaller circles. Arrows show flows of knowledge from one source to another, with the thickness of the arrow proportional to the number of times communication of research findings occurred between sources.

Where possible, identify 'boundary organisations' that are able to cross boundaries between otherwise disconnected networks of actors, including researchers and policy-makers. There will often be key individuals within these organisations who understand the research and are well connected to and trusted by the policy community and who may be able to help you engage credibly with their contacts. If you understand who has influence, you can start to identify the messages from your research that are likely to resonate with these influencers and develop a communication strategy that will enable you to build relationships with these key people, who will then open doors to the policy community for you.

A few years ago, I decided to try and trace how research was getting into policy and practice (or not), and my colleagues and I chose 77 different research findings and traced how they travelled from peer-reviewed literature into policy and practice through social networks using social network analysis and interviews with those who had found out about the research, to see how they learned about it and who they had passed it on to. One of those findings was the work on peatland carbon that I'd been involved with, and Figure 15 is a simplified representation of the network map showing how that research got into policy and practice. It shows how peatland researchers tend to mainly communicate their findings through scientific journals, which are not used directly as a major source of knowledge by policy-makers or those who seek to influence them. On the other hand, researchers in this case were as good at communicating their findings to NGOs and charities as to policy-makers directly, and it was through these NGOs and charities that most of the information reached the government.

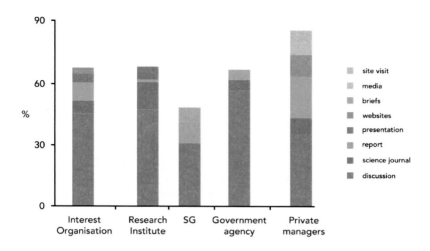

Figure 16: Case study research showing how different groups find out about new research findings, based on interviews in Scotland as part of the Ecocycles project

This presents an interesting dilemma for researchers. Charities and lobby groups have more time and resources to promote research findings that support their causes than researchers typically have, but they have an incentive to present a selective or biased representation of the research. Again, this often comes down to relationships. Although it is impossible to control how others represent our research, by engaging with these knowledge brokers, it is possible to increase the likelihood that they fully understand it, including important nuances, caveats and remaining uncertainties. If you can create a strong, trusting relationship with key people in these groups, they are more likely to keep you informed of the way they are using your work, and respond proactively if you spot problems with the way they are using it.

How to build relationships with policy-makers

Relationships are at the core of my approach to working with policy-makers. My own research and other published evidence shows that although policy-makers find out about research from many sources, it is information from face-to-face contact with people they trust that most commonly influences decisions (Figure 16). It doesn't matter whether it is at a one-to-one meeting or at a workshop, conference or seminar, and it doesn't matter if the contact is directly with the researcher or more indirectly via some sort of intermediary, for example, someone from a government agency or a charity or lobbying group. The important thing for researchers is to invest time in developing trusting, two-way relationships with key members of the policy community working in their field.

Although Figure 16 is based on just one case study, the key points appear again and again in the literature: policy-makers find out about much of the research they use through face-to-face contact with trusted sources. The graph above shows that although the interest groups and agencies advising and lobbying the government do use journal papers as a source of knowledge, they are far outweighed by face-to-face communication with trusted sources. While they do use policy briefs, again, face-to-face discussion with trusted sources is the most important way they find out about research evidence. Creating a policy brief is not enough: it is what you do with your policy brief that counts. Leaving a policy brief as a reminder of key points and a link to further information after a face-to-face meeting with someone is far more likely to effect change than simply mailing out briefs and hoping someone reads them.

Finally, it is important to consider how you will demonstrate the credibility of your message, when you're not going to have time to present all the methods and data that lie behind it. In many cases, this credibility can be earned by proxy, by referencing a key paper in a prestigious journal that your findings are based on, and by being introduced under the brand of your funders or by key figures who are already trusted by the policy community.

Influencing policy

In the world of politics, emotion is often used to bias decisions away from the evidence. Many researchers prize their objectivity and detached independence. However, positive, effectively channelled emotion gets people's attention — it makes people sit up and take notice. Using emotion appropriately, as researchers, we can connect with our audience and engender empathy. By engaging with both hearts and minds, we increase the likelihood that our audience is really listening, and actively considering how our evidence fits with the other evidence they have access to, their goals and their worldview.

As experts, we hold a privileged position of authority, which is likely to be a key factor in getting us an audience with decision-makers in the first place. Retaining that credibility is essential, and so it is important to carefully channel our emotions. Decision-makers are more likely to respond to positive emotion than they are to anger or doom and gloom predictions. For example, there is evidence that decision-makers are less receptive to messages about the value of nature if these messages are perceived as threatening their psychological needs of autonomy (e.g. because they feel manipulated or coerced), happiness (e.g. environmental and sexual health campaigns based on fear), reputation (e.g. because they feel implicitly criticised or patronised) and self-esteem (e.g. because they start to feel responsible for or guilty about the issues concerned). On the other hand, enthusiasm is infectious. Presenting our evidence with passion and crafting our arguments to meet the innate psychological needs of our audience is more likely to get people to listen, even if they don't act on what they hear.

So far it has been implicit that the evidence itself is unquestionable; the question is only whether we as researchers should use emotion to communicate that evidence. But there are many researchers who would challenge the idea that research is (or can ever be) entirely independent, emotionally detached and objective. If we recognise that our research is just one strand of evidence feeding into what are usually political decision-making processes, then we can begin to explore the subjectivity inherent in many of the processes we use to generate 'evidence', and can become less detached and more emotionally engaged in the normative goals that pervade our work.

Traditionally, the mass media has been an important way of amplifying messages, so policy-makers receive our messages from many different sources and are given a sense of the weight of public opinion behind that message. Nowadays, social media is an increasingly important way of amplifying messages in a more relational way, raising awareness of the issues we're working on and demonstrating wider public support for our ideas very transparently via counts of retweets, likes and views. Whether you're using social media or not, developing a partnership or 'policy network' with other individuals and organisations who are interested in the messages arising from your work can also help amplify your message, getting your research to different members of the policy community in different ways (these approaches are considered in more detail in the previous section). Importantly, it is often possible to directly engage with members of the policy community around these messages via social media, enabling you to persuade rather than simply using the pressure of the mass media. Of course, in many cases, having visibility in both mass and social media can further amplify your message.

Practical tips for influencing policy through relationships:

1. **Develop a structured and systematic engagement strategy:** it doesn't have to be written down; even if it is only in your head, thinking systematically about how you will engage with key stakeholders can significantly improve your chances of being relevant and helpful. First of all, map your stakeholders to work out who actually holds decision-making power within an organisation. Often, people with high levels of personal and transpersonal power have greater ability to actually make things happen in an organisation than the people at the top of the hierarchy. Also beware of automatically gravitating towards the 'usual suspects' who are highly visible, and consider whether there are marginalised and powerless individuals or groups that could really benefit from your work, and who may be highly motivated to work with you. Once you've worked out who you want to engage with, you need to work out what's likely to motivate them to engage with you. What messages from your research might resonate with their interests and agendas? What modes of communication are they most comfortable with? What is the best timing/occasion for communication? What sort of language do they use or avoid? If you can't reach those with decision-making power to start with, identify people in the organisation who are more likely to engage with you, and expand your network from there. Those with decision-making power are more likely to listen to you if the rest of their team are already listening to you.
2. **Empathise**: put yourself in the other person's shoes; work out what motivates them, how they might be feeling, and what they might want from your research. Work out what is likely to build trust in the relationship between you. For example, do you need a letter-headed initial approach or do you need to be introduced over a pint of beer? It may be worth doing some

digging about the person and their organisation to help you empathise effectively. For example, you might research them online, looking at their profile and the sorts of things they're writing or tweeting about. Alternatively, you might ask what others in your network know about them. If none of these ideas work, you can talk to other people in similar roles first, to get a feel for the sorts of issues that are likely to motivate them, and the language and modes of communication they're likely to respond best to. It's a bit like the sort of process an actor or actress would go through to research a role they've been given.

3. **Practise your communication skills**: find out about body language, so you can read how the people you want to influence are reacting to you and adjust your approach accordingly. Adopt confident body language. If possible, practise in front of a camera or mirror, or get feedback from a colleague. You need to practise open body language that tells the other person you trust them and that they can trust you (e.g. avoid crossed arms and legs and give plenty of eye contact). Think about your handshake and what it conveys — a firm handshake conveys confidence and is more likely to instil trust than a limp one. Put your pen down when you're not writing and make sure there are no physical barriers between you (e.g. a pad of paper propped up between you). Be mentally aware of your facial expressions, to make sure you're not slipping into a scowl as you concentrate on what the other person is saying; try and be as smiley as comes naturally to you. It is important to make it clear you're listening and genuinely valuing what they tell you with nods and non-verbal, encouraging sounds. If you're really listening with all your heart, you'll find yourself naturally mirroring the other person to an extent. For example, if you make a strident start and discover the other person is very quiet and shyly spoken, you'll probably feel uncomfortable continuing to talk loudly and confidently, and will moderate your behaviour to be less different from them. If you are able to adapt your tone of voice and body language to theirs, they are likely to feel respected and more able to connect with you. If all of this doesn't come naturally, start small and build from there. Like other roles you have to adopt professionally (e.g. lecturing), with practice it will become second nature, and eventually become entirely natural.

4. **Give**: ensure there genuinely is something in the engagement for the other person that they really want, and think about how

you'll deliver those benefits in concrete terms in the near future. If you've managed to really empathise with them, then this bit should be easy.

5. **Assess your power** in the context of the stakeholders you want to work with, bearing in mind that you may be significantly more or less powerful in different contexts (see Box 10). For example, in some contexts, as a result of negative experience with other researchers, your status as an academic might mean people expect you to be irrelevant or exploitative. Whereas, with a different group of stakeholders, your status as an academic might mean your view carries greater weight. One way of thinking about how powerful you might be in a particular context is to think about your levels of:

   - Situational power (e.g. your level in formal hierarchies, access to decision-makers)
   - Social power (e.g. your social standing, race, marital status or whether you have Dr or Professor in front of your name)
   - Personal power (e.g. how charismatic, trustworthy and empathetic you are perceived to be)
   - Transpersonal power (e.g. a connection to something larger than yourself, ability to transcend past hurts, freedom from fear and commitment to an altruistic vision)
   - If you don't have enough power or legitimacy yourself, then think about ways you might be able to improve your personal and transpersonal power (as these are easier to change than your social and situational power). And if you need a quick shortcut to more power, get yourself introduced or be accompanied by someone who is already well trusted and perceived to be legitimate in the eyes of the people you want to work with. You can assess your own levels of power using the prompts in Box 10.

6. Finally, where necessary, **go around or above obstructive individuals**, developing a tailored engagement strategy for the next person in the organisation you need to engage. If you have done everything you can to adapt to the needs and priorities of someone who is preventing you from reaching those with decision-making power in an organisation, see if you can find others in the organisation who have slightly different needs and priorities, or a different world view or perception of risk, and see if they will open a door to decision-makers on your behalf instead. Some individuals are naturally more likely to be

receptive to new ideas (they are 'early adopters' of innovations) and others (sometimes termed 'laggards') hang back and wait for others to try new ideas first. You need to identify the innovators in the organisation you're trying to influence. In a few rare situations, there may be a case for going to someone higher in the hierarchy. For example, a minister might see a political opportunity in a high-risk idea emerging from research that their civil servants might not have been willing to consider. However, under instruction from the minister, civil servants are likely to be happy to investigate your ideas.

Sustaining trusting relationships

Where possible, get feedback on your interactions with policy-makers, whether directly (e.g. via feedback forms after a workshop) or indirectly from colleagues' observations of your interactions. Seek out colleagues and peers from your discipline who are already successfully working with policy-makers on related issues, and learn from them. They will be able to advise on the key people to communicate with, and how best to approach each person. If possible, take opportunities to watch colleagues who have experience of collaborating with policy-makers at work. You can also study examples of successful (and unsuccessful) policy uptake of similar research. By understanding the factors that led to (or prevented) policy uptake, you may then be able to identify mechanisms you can use or avoid yourself. Although research project funding typically stops after between three and five years, it is important to find ways to sustain engagement with the policy

community long after a project ends if your research is to really make an impact. Only with sustained engagement is it possible to develop trust. It is these trusting relationships that will get you the ear of policy-makers, and enable you to adapt your research to their needs. Finally, be tenacious: put the same effort into building relationships with the latest civil servants to move into the roles you need to work with, again and again...

Top pathways to policy according to researchers

I'd like to conclude with a perspective from researchers who have worked with policy-makers on the key pathways that enabled them to effect change. In 2015, the Higher Education Funding Council for England published a searchable database of impact case studies, collected as part of its evaluation of UK research under their Research Excellence Framework. I commissioned an analysis of a 5% sample of impacts on social (including health), economic and environmental policy, and classified the different pathways to impact that the researchers identified (you can read the full results in the table below). The most commonly cited impact pathways make interesting reading:

1. **Publications**: as you might expect, the number one pathway described by researchers in their case studies was academic publications, typically in peer-reviewed journals. Given that most case studies contained multiple pathways, the role that academic publications played in achieving impact is debatable. Also in the top ten pathways, however, were industry publications and policy briefs, underlining the importance of translating academic findings into formats that are more likely to be read by non-academics.

2. **Advisory roles**: being asked to contribute to government inquiries, reports, panels and committees was one of the most important ways that researchers influenced policy, with over 50% of the case studies we reviewed using this pathway.

3. **Media coverage**: researchers perceived that getting their research covered in the mass media was an important route to policy impact. This might be because of the visibility that media coverage can afford research, putting it directly in front of decision-makers who engage with the media (themselves or

because they are made aware of media coverage by their civil servants), or more indirectly by contributing to a body of public opinion that decision-makers then respond to.

4. **Partnerships and collaborations** with industry and NGOs: by finding organisations that shared their research interests, researchers may have been able to harness the lobbying power of these organisations to promote their work more actively and at higher levels than they would have had the time, resources and ability to do as researchers on their own. These partnerships also enabled researchers to test their research in real-life situations, which gave it more credibility when approaching policy-makers.

5. **Presentations** with industry, the public and government: face-to-face meetings, whether one-to-one or in workshops and conferences, can be a powerful way to get research findings noticed and understood, partly because the audience has the opportunity to question the research team. Although researchers cited presentations directly to government, they were just as likely to cite presentations to industry and the public as their pathway to policy impacts. This may suggest, like the previous point, that many impacts were achieved via the knowledge brokerage role of industry partners, or by raising the public profile and contributing towards a weight of public opinion that policy-makers could not ignore.

6. **Developing easily accessible online materials** based on the research was also a commonly cited pathway to policy impact. Although this is rapidly changing with open access, a significant proportion of research findings (particularly older material) is behind journal paywalls. Making this material both available and easily accessible via online materials that translate the findings for specific audiences can be an important way of getting research into policy.

One of the most important pathways was advisory roles. Although these roles are sometimes one-off interactions, for example, giving evidence to a parliamentary committee, many are medium- to long-term roles over a period of years, in which the researchers are able to build trust with other panel/committee members and provide advice on an ad hoc basis between formal meetings. Apart from this, however, the majority of the main pathways were in dissemination mode. Although partnerships and collaborations with industry and NGOs feature strongly in the pathways reported by researchers in these case studies, these organisations appear to primarily have operated as knowledge brokers, helping to translate and amplify messages arising from the research, and enabling them to reach policy-makers.

It is clear from the case studies we reviewed that well-targeted dissemination of research findings can pay dividends, but if the experience of these researchers is anything to go by, certain types of dissemination may be more likely to achieve impact, for example, publications, online resources, press releases and presentations. But it is just as important to invest in longer-term relationships with the policy community and key players who have the time, resources and expertise to help you form those relationships and amplify your message. In the long term, this may open up opportunities to contribute to advisory committees and other processes that directly feed into the policy process.

Chapter 21

How to make a policy brief that has real impact

Have you ever wondered if the policy briefs you've produced actually made a difference? There are many guides that will tell you how to write an effective policy brief, but is the wording and design what makes the difference? Well, partly. If you want to make an impact, writing the brief is just a small part of the work. A policy brief is only worth what you do with it.

If you want to take a relational approach to developing your next policy brief, you need to consider how you engage members of the policy community in your design and planning, writing, distribution and longer-term engagement.

1. Design and planning

Ask yourself the following questions to put policy-makers at the heart of your design and planning:

1. **Who is the policy-maker?** This is important because it determines the target group of the policy brief. Are you targeting people within specific government agencies, who are likely to have a relatively focused interest in the topic, with a relatively high degree of technical competence? In this case, you will need to include some of the technical detail, so that these specialists can make up their own mind about the credibility of your work. Or are you briefing policy analysts within government departments who advise ministers, or the MPs and ministers themselves? In this case, your policy brief should be much shorter, with far less technical detail and much simpler language.
2. **When are they likely to read a policy brief?** This might determine when and how and in what format to distribute the brief (e.g. electronic or paper version, when to schedule the email with the brief attached, such as an evening, even on weekends, will it be read over breakfast or on a train/flight?)

3. **How much time do they have to brief themselves on the latest research?** This is crucial for deciding the length of the 'brief'. If you are a high-ranking politician, you may only want to read a single page. Others might spend up to 30–60 minutes to get a more detailed picture of the research behind your recommendations. One approach is to do a 'breakfast test': can your policy brief be read and understood in the length of time it takes to drink a coffee over breakfast?
4. **Why should they pick up the policy brief in the first place?** What is likely to grab their attention? How can you make it visually attractive, with a heading that is of interest? What sort of 'strapline' or 'teaser', perhaps based on a key finding, might encourage the politician to read it?
5. **What do they want to know?** What are the most pressing, wider policy issues? Can you link to important and current policy questions and issues? If your work is only one small contribution to a wider issue, can you collaborate with other researchers working in the same area to create a policy brief that includes your research, but that is likely to be perceived as having greater political significance? Is now the right time to put out your policy brief if there are other major issues swamping the policy agenda in your area?
6. **Is this compatible with their overarching goals and ideology as a policy-maker?** Many policy-makers are looking for research that furthers their own agenda and legitimises their views and ideology. They are unlikely to change these fundamental values and beliefs on the basis of one policy brief, so make sure you phrase your recommendations carefully to avoid provoking a negative reaction based on a presumption of ideological incompatibility. This doesn't mean you need to make political recommendations or change your findings to fit the views of politicians — far from it. It is surprising, however, how far you can adapt the way in which you communicate your findings to make them attractive to different policy actors without altering the research in any way.
7. **What reason do they have to trust you?** What indicators can quickly reassure a policy-maker with limited time that you are knowledgeable and credible enough to deliver the message? If you do not have a high profile yourself, what indicators of esteem might make them trust you by proxy, such as your institutional affiliation, the badge of your research funder or

more senior academic mentors and supervisors who helped you write the brief?
8. **Are there clear and actionable things they can do as a result of reading this?** Is the evidence you provide aligned with the policy problem that the policy-maker needs to address? Can you provide solutions to these problems? Are your recommendations SMART (specific, measurable, achievable, relevant, and time-bound)? Can you make it even SMARTER i.e. 'effective' (e.g. cost-effective) and therefore more 'realistic'?

2. Writing and stress-testing

Now you have put yourself, effectively, in the shoes of your policy audience, you need to ask yourself:

- What would you personally like to get across?
- What's your own aim for the policy brief?
- Does that match the policy-maker's perspective?

If your answer to the last question is "no", you should stop right there, otherwise you might be wasting your time.

If, however, you have been able to align your aims with the needs of policy-makers, it is now time to write the brief. With the help of the questions above, you will already have decided on length, style and language. You are using common terms without too much jargon, and avoiding (or, if you can't, spelling out) acronyms. You are telling a convincing story about why change is needed.

How to set up the brief itself?

On the front page you'll need:

- **Title**: keep it short and powerful — would you personally pick up a policy brief with such a title? You can consider adding a subtitle, if it further explains your main message (again keep it short).
- **Teaser**: start with a summary of the brief's content and its relevance in two to three sentences (maximum five lines), state all the main points and repeat them throughout the document.
- **Recommendations**: in bullet points, perhaps use a sidebar or box.

- **Picture/photograph**: something attractive and positive that captures the research topic well. Make your picture bigger and have less text if possible.

On the next pages, consider the following:

- **Overview**: give a brief overview and state the problem or objective. Embed your research in an important, current issue and explain how the policy brief contributes to that issue and provides useful answers.
- **Introduction**: summarise the issue, explain the context (including the political) to explain why the topic is so important and how your research can help to solve/improve the situation. Pinpoint gaps in current policy, link to crisis points that may be windows of opportunity in which new policies may be looked for. Outline a brief history or background, but only if it is relevant to the theme (otherwise leave it out!).
- **Research findings**: these are the answers from your research that help to solve the problem (other findings may be of interest to researchers and might look pretty on a graph, but if they don't help address the policy issue, cut them out). If possible, present your findings in a more visual, clear style, so the idea can be grasped immediately. Include research evidence from the literature and other sources to support your own findings in plain language. Use subheadings to break up blocks of text (keep sections of text and paragraphs as short as possible). Any graphs or other figures should be simple, and be labelled with a short description that can be understood without reading the text.
- **Sidebars and boxes**: highlight the most important evidence in sidebars or boxes, so people can easily skim through the key points if they are in a hurry (remember these are for highlighting important things, not for unimportant things, to policy-makers at least, like definitions).
- **White space and photographs**: try and break up your text with plenty of white space and photographs to avoid intimidating readers and also to make your work more attractive to engage with. If you can, hire a professional designer to help with this. If there's not enough room to fit everything in that you want, don't make the font size smaller or cut white space and images — cut down your material (the next stage in the process, the feedback

loop, will help with this if you're struggling to work out what you can cut).

- **Additional sources**: more (background) information, more detail on the topic, maximum four further sources, including peer-reviewed material by you and your team

Last page:

- Brief summary statement, concluding with the **take-home message**
- **Policy recommendations**: clear recommendations aimed at a specific policy sector (or sectors) and specific live policy issues, in bullet points, stating why these options are recommended
- **Author's contact details**: including current position, associated institute and funder (remember the credibility issue), Twitter accounts (for key project staff and the project itself if this account exists), websites etc.
- **Acknowledgements:** if necessary (e.g. your funder)
- **Sources**: cite in footnotes, if needed

Stress-testing

If you want to take a relational approach to developing your policy brief, the next step is to stress-test it. I usually move from low to high stress-testing, starting by sending a draft of my policy brief to trusted colleagues who have not been involved in its production, before sending to members of the policy community who I think are likely to hold very different views on the issues I'm writing about:

- **Academic content:** I will start by stress-testing my content with other researchers. Do they agree with my interpretation of the evidence? Have I missed any important evidence? Could I communicate uncertainty more effectively?
- **Design:** if I have designed it myself, I will send it to a few friendly colleagues for comment on the layout and selection of photos.
- **Language:** I will send it to a science communication specialist or a non-academic friend to get feedback on my choice of language. If they cannot understand my message, then I will try and rewrite it. If I am aiming for EU policy-makers (most of whom are non-English native speakers), I try to give my draft to non-native English-speaking friends for feedback. Alternatively,

identify jargon using the Up Goer Six website (http://www.splasho.com/upgoer6/), a text editor that colour codes all words according to how common they are.

- **Messaging:** finally, I will seek feedback on how the evidence-based messages in my policy brief are coming across to different audiences within the policy community. For a controversial topic, I will specifically seek opinions from people who I know hold opposing views. What are the weaknesses and limitations? If you were in a debate with me, what holes would you pick in my argument? Based on this, I can now predict some of the objections that might arise when I start trying to communicate my message more widely. In some cases there is little I can do to make my message resonate with different sides of the debate, but at least I know some of the questions I am likely to get. However, in other cases, it is possible to plug gaps and strengthen or reframe arguments. If a fundamental flaw in your argument is revealed, or you are pointed to contradictory evidence, you have time to correct your mistakes.

3. Distribution

How should you distribute your policy brief? The options are growing rapidly:

- **Electronically**: first you might upload your brand new policy brief to your own and your department/organisation's website. This will provide you with a link to a PDF of the brief that you can include in emails that you send out to your target group.
- **Hard copy**: sending a 'paper' version to your target audience is important. Do not just send to a department, but make it personal and send it directly to a person. Even better, you can hand over your brief in person to the policy-maker in a face-to-face meeting (be it over lunch, at a conference, during their 'office hours' — this might depend on your previous attempts to start a relationship with your target audience).
- **Social media and beyond**: use the PDF link you created for all social media that you have set up personally and within your team, organisation, department or institution. That may (for example) include Twitter, ResearchGate, LinkedIn and even Facebook. Use a picture/photo of the cover (or key photo) of the brief to accompany distribution via social media as this

attracts people and increases the likelihood of further distribution by sharing (liking, retweeting etc.) by others in your network. Make sure your profile on social media is consistent with your role as an expert in the field, with a link to your institution or a webpage that clearly links to it. The more times your target audience comes into contact with your material via different channels and people in their network, the more likely they are to perceive that it must be worth engaging with. For this reason, you might also ask your PR department if they can publish a press release (together with the original research paper/research on which the brief is based) on Twitter and so on. Furthermore, consider writing a blog post about the brief that includes the recommendations, and distribute it through the channels mentioned above.

4. Engagement and impact

Follow up the email to your targeted people with a phone call. Ask if any further information is needed. Propose a lunchtime meeting or seminar to discuss your research further. Make sure the brief remains in the memory of your target group beyond the mere picking up and reading of it. You can also invite them to related conferences and workshops and take a copy of the brief with you to any of these events. Remember that one-way information flows are unlikely to get anyone to act on your recommendations.

If you are not likely to meet the target of your policy brief any time soon, you might start following them via Twitter (as mentioned earlier, lots of policy-makers are active on this platform nowadays) or subscribe to email lists to know what they are up to and to learn where your work fits in with and contributes towards their agenda. Take the time to find out what they think, what sort of language they use, what is on their agenda and how you can help them with their daily tasks. And when you have the chance to meet them, your connection via social media will make it easier to build trust.

Perhaps you will find forming trusting relationships so fruitful that you decide to co-produce the policy briefs in collaboration with the people who will use it. This is a particularly effective way to develop the policy brief according to their needs and will ensure that it is used and result in impact.

To be able to achieve impact, the best-case scenario is that you already have a long-lasting, trusting relationship with relevant policy-makers. But it is not too late; you can start now. Find out which events they are likely to attend, and look up photographs of them, so that you can identify them during breaks to introduce yourself to them and get to know them. Policy-makers are just people like us. If you find it difficult to start small talk by yourself, ask colleagues to help. They may already be trusted by the policy-maker and may be able to introduce you to them. Some of this trust will make your initial contact more trustworthy too.

Examples

Finally, I'd like to show you a few examples of policy briefs that I think are particularly good. The first was developed by Julia McMorrow from the University of Manchester, and is notable because it led to concrete changes in government policy. It raised cross-sector awareness of wildfire and helped make the case for severe wildfire to be included for the first time on the National Risk Register in 2013. The Chair of the Chief Fire Officers Wildfire Group commented:

"Such was the quality of the Policy Brief, that I used it to raise the awareness of wildfire issues affecting UK Fire and Rescue Services by circulating it to all Chief Fire Officers... The work is as relevant now as it was when first produced in 2010. The FIRES Policy Brief also formed a cornerstone of the Wildfire Group's initial Action Plan."

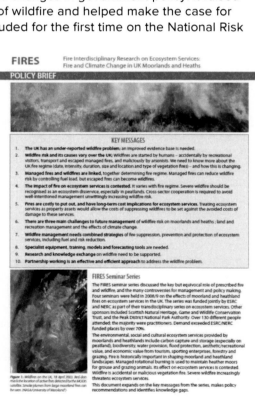

The policy brief recommended better fire reporting and as a result Julia was invited to work with the Fire Service to evaluate how satellite data and their Incident Recording System could be used to understand national and regional wildfire distribution. The joint research developed criteria to differentiate 'wildfires' from other less significant vegetation fires and recommended ways to improve reporting. The definition was used in the Scottish Government's Wildfire Operational Guidance. The work has also been used as an example to influence wildfire policy in Ireland. Julia was invited to join the England and Wales Wildfire Forum, the Fire and Statistics User Group and other national and regional stakeholder groups.

I asked Julia what she thought had made it such an effective policy brief, and she explained the long path that she and her colleagues took to develop it. First, she organised a series of seminars, to which she invited all the key stakeholders who were affected by the issues she and others were researching. Part of this was about presenting and discussing her research findings, but it was also about understanding how different stakeholders perceived the research, and appreciating their knowledge of the issues too. She ensured that the steering group of the seminar series was composed equally of practitioners and researchers. They jointly took the policy brief forward, deciding on the language to be used, and the framing of the key messages, ensuring all the time that it remained based firmly on the seminars' findings. Part of the group was an organisation who had already run a successful series of briefing notes on related topics, so their design template was used to reach their existing audience and make it as widely accessible as possible. Julia explained:

"The most rewarding part of developing this policy brief was the relationships we built leading up to and during the process, which have stood the test of time. It also opened doors to influential national stakeholder groups. In both these ways, it continues to bring us new opportunities to realise impacts from our research."

For me, this is a really powerful example of the relational approach to developing policy briefs I've described in this chapter. The priority of the team was on building long-term, two-way, trusting relationships through a series of meetings, which enabled them to co-produce the text. Whatever design ideas the team might have had were put aside, so that an existing, well-recognised design template could be used. This enabled the team to make the material as widely available as possible. After the policy brief was published, the research team was able to continue working closely with the members of the practitioner and policy community who had been involved in the seminar series to effect policy change.

Figure 17: Examples of policy briefs from the Evidence Matters and NIEER series

The University of **Nottingham**
UNITED KINGDOM · CHINA · MALAYSIA

NHS

CLAHRC **BITE**

A bite-sized summary of a piece of **CLAHRC** research

January 2013
BITE 20

Who this applies to:

Practitioners prescribing for patients with moderate-to-severe Alzheimer's disease.

Findings and implications...

- Cholinesterase inhibitors were effective for severe Alzheimer's disease.

- These findings are inconsistent with the literature on which NICE guidance was based (due to be updated in 2014) which recommends AChE inhibitors, such as donepezil, for mild to moderate Alzheimer's disease and memantine for severe disease.

- This study recommends treatment with cholinesterase inhibitor should be considered in the awareness of the increased risks of adverse outcomes including syncope, the need for permanent pacemakers and hip fractures.

While donepezil is currently recommended for patients with mild or moderate Alzheimer's, recent data indicates that donepezil gives cognitive and functional benefits for patients with moderate or severe disease.

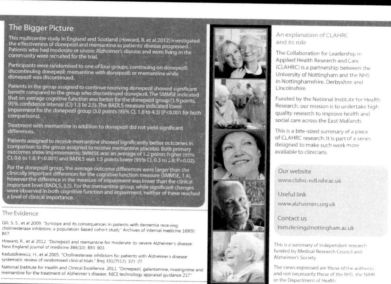

The Bigger Picture

This multicentre study in England and Scotland (Howard, R. et al.2012) investigated the effectiveness of donepezil and memantine as patients' disease progressed. Patients who had moderate or severe Alzheimer's disease and were living in the community were recruited for the trial.

Participants were randomised to one of four groups: continuing on donepezil; discontinuing donepezil; memantine with donepezil; or memantine while donepezil was discontinued.

Patients in the group assigned to continue receiving donepezil showed significant benefit compared to the group who discontinued donepezil. The SMMSE indicated that on average cognitive function was better for the donepezil group (1.9 points, 95% confidence interval (CI) 1.3 to 2.5). The BADLS measure indicated lower impairment for the donepezil group (3.0 points (95% CI, 1.8 to 4.3) (P<0.001 for both comparisons).

Treatment with memantine in addition to donepezil did not yield significant differences.

Patients assigned to receive memantine showed significantly better outcomes in comparison to the group assigned to receive memantine placebo. Both primary outcomes show improvements: SMMSE was an average of 1.2 points higher (95% CI, 0.6 to 1.8; P<0.001) and BADLS was 1.5 points lower (95% CI, 0.3 to 2.8; P=0.02).

For the donepezil group, the average outcome differences were larger than the clinically important differences for the cognitive function measure (SMMSE, 1.4), however the difference in the measure of impairment was lower than the clinical important level (BADLS, 3.5). For the memantine group, while significant changes were observed in both cognitive function and impairment, neither of these reached a level of clinical importance.

The Evidence

Gill, S. S. et al 2009. "Syncope and its consequences in patients with dementia receiving cholinesterase inhibitors: a population-based cohort study." Archives of internal medicine 169(9): 867

Howard, R. et al 2012. "Donepezil and memantine for moderate-to-severe Alzheimer's disease." New England journal of medicine 366(10): 893-903

Kaduszkiewicz. H., et al 2005. "Cholinesterase inhibitors for patients with Alzheimer's disease: systematic review of randomised clinical trials." Bmj 331(7512): 321-27

National Institute for Health and Clinical Excellence. 2011. "Donepezil, galantamine, rivastigmine and memantine for the treatment of Alzheimer's disease. NICE technology appraisal guidance 217."

An explanation of CLAHRC and its role

The Collaboration for Leadership in Applied Health Research and Care (CLAHRC) is a partnership between the University of Nottingham and the NHS in Nottinghamshire. Derbyshire and Lincolnshire.

Funded by the National Institute for Health Research, our mission is to undertake high quality research to improve health and social care across the East Midlands.

This is a bite-sized summary of a piece of CLAHRC research. It is part of a series designed to make such work more available to clinicians.

Our website
www.clahrc-ndl.nihr.ac.uk

Useful link
www.alzheimers.org.uk

Contact us
tom.dening@nottingham.ac.uk

This is a summary of independent research funded by Medical Research Council and Alzheimer's Society

The views expressed are those of the author(s) and not necessarily those of the NHS, the NIHR or the Department of Health.

Figure 18: Example of a CLAHRC BITE

Figure 19: Example policy brief from the Living With Environmental Change policy brief series (pages 1–4, clockwise from top right)

Finding attractively designed policy briefs is remarkably hard. However, the National Institute for Early Education Research (NIEER) have a highly visual format to their policy brief series, which I love (Figure 17). Their briefs are full-colour throughout with background colours selected to match colours in the photographs that feature on each page. As you can see from the front page, policy recommendations are clearly identified and highlighted here, along with a summary of the literature on the topic (not just the narrow findings of one particular study). The Evidence Matters series is similarly colourful, featuring full-colour photographs on the front page (Figure 17). This series operates like a magazine, with monthly briefings on a specific issue. Having regular releases of new policy briefs helps raise the profile of a series, keeping copies regularly at the top of the pile on coffee tables in the offices of those you want to reach out to. The CLAHRC BITEs series is also colourful, featuring the National Health Service (NHS) logo which is widely recognised in the UK (Figure 18). It is a great example of what can be done with a short format. These bite-sized summaries of evidence are only two sides of A5 paper, but they convey the evidence concisely and powerfully.

In contrast to these, the Living With Environmental Change policy briefs (Figure 19) are monochrome green, but this was done for a clear reason, as Anne Liddon, the series editor, explained to me:

"We launched a similar series ten years ago with the research councils' Rural Economy Land Use programme (RELU) which had a slightly different focus. The RELU policy briefs were incredibly successful and we gained a reputation for providing timely and relevant research findings to Government and other stakeholders. We worked hard on the RELU brand, and the policy briefs were instantly recognisable as part of the programme. So when RELU ended and Living With Environmental Change wanted to launch a new series, we managed to merge the branding so the new series kept the same look and format. This meant policy-makers instantly recognised and trusted the new series as a reputable source of information that could inform their decisions."

One of the things that is interesting about this is the importance of brand reputation and recognition for policy briefs. You can just create your own design template and do your own thing. However, if you can find an existing policy brief series that has already built a relevant audience and has a strong reputation, your policy recommendations are more likely to be read and paid attention. If you are starting a new policy brief series, work on your brand and create something distinctive, attractive and instantly recognisable.

The now discontinued Living With Environmental Change series is a great example of what a good policy brief can look like, particularly on the inside pages. As series editor, Anne encouraged researchers to focus on specific key findings, rather than covering the whole research project, with a strong emphasis on the implications for policy. The front page has an image that tries to capture the content of the brief. In this example, it took a long time to find an image for air quality. The researcher wanted a positive image of air quality, so pictures of traffic and exhaust pipes were out. However, the researcher's suggestion of a landscape image did not seem obvious enough. Eventually, Anne sourced an image of a colleague's daughter running with a kite, and the search was over. On the inner pages, there is always plenty of white (or green) space around the text, no matter how much the researcher pleads to add more words. The introductory paragraph lays out the problem being addressed, and each heading is a question that Anne thinks the audience will want to ask. Finally, there is a box section with concrete action points for the audience and further information.

There are many more examples I could show you, but these four give you a flavour of the sort of thing that is possible. I've chosen them because they provide important lessons that illustrate and complement the suggestions I've made earlier in the chapter. However, take a look around for yourself at policy briefs, whether or not they are linked to your research area, and draw on the best ideas.

Chapter 22
Tracking, evaluating and evidencing impact

How do you know if your research actually made a difference?

If you think you made a difference, can you prove it?

In this penultimate chapter, I want to consider how you can track and evidence the impacts that the rest of this book is designed to help you generate. First, I want to tackle the challenge of motivating yourself to keep track of your impacts as they arise. This can make your life much easier when you are asked by research funders and others to report on the impact of your work, but the time involved puts most researchers off. I have developed an approach that takes me less than a second to log impacts in a place where I can find them later. In the next section I explain my approach, but I want to encourage you to find your own approach.

Tracking research impacts efficiently as you go

Many universities now have online systems on which they ask researchers to log impacts. Although these systems are powerful, and increasingly essential for managing the assessment of impact across large institutions, the majority of academics do not regularly engage with them.

As a researcher myself, I understand the challenge. I told my research funders about my impacts last week via their annual request to input data to their online system. The week before, one of my funders contacted me to write an article about my impact for their magazine. Every couple of months I have an informal call with another of my funders and tell them about the impacts of the project they are funding. Every quarter I have to write a report on my research impact for the programme that funds my Chair position. I'm drafting an impact case study for the next Research Excellence Framework. On top of this, my university also expects me to log my impacts in its research management system (it is on my to-do list).

I suspect that I am victim of my own success, as few of my colleagues have to report impact as much as I do. However, most researchers I know agree that the administrative burden of reporting impact is rapidly ballooning.

This is a problem for two reasons. First, evidencing impact is important for universities' reputations and bottom lines. Based on an analysis of the UK's Research Excellence Framework in 2014 (that I published in the Fast Track Impact magazine with Simon Kerridge from the University of Kent), a top-scoring impact case study was worth £324,000 on average over seven years, and it will be worth more in the next assessment in 2021. Second, few researchers keep records of their impacts as they occur, and so tend to rely on memory when they report impacts (usually in a rush, just before the deadline). This means reporting is often incomplete and lacking detail, leading to missed opportunities to deepen or properly evidence impact.

For me, this is a problem of hearts and minds. Most researchers' hearts aren't in impact reporting. They would prefer to be generating impact rather than entering it into an online system that they've forgotten how to access or navigate. Even if your heart is in it (as mine is), the repeated requests to input different impacts to different people in different ways are likely to become increasingly frustrating.

What we need is a culture of tracking impacts as we go, and the only way we can create this culture is by going for both hearts and minds. For me, the easiest way to get to the heart of impact tracking is to link it to your impact goals. I use my Fast Track Impact Tracking Template (Table 3) to get researchers to visualise their impact goals (working back if necessary from the people who are interested in their work and why they are likely to be interested). It can be motivational to visualise the impacts your research might have, imagining yourself years from now, looking at what has changed because of your work.

However, I tell my colleagues that there is no point in visualising their future impact if they have no way of telling whether or not they are moving towards or away from it. By looking around themselves, in their mind's eye, they can start identifying the specific things that have changed, that tell them they have reached the impact they set out to achieve. Now, rather than just waiting for evidence to appear, they are looking for specific things, and measuring them on a regular basis to check if their impacts are on track. If they are off track, they are empowered to change their pathway to impact in ways that are likely to get them back on course. Impact tracking now has a purpose. It isn't just filling in forms to keep some nameless bean counter happy. It is actually increasing the likelihood that they achieve the impacts they want to see. From this place of inspiration and empowerment, I'm motivated to track my impacts. We've got to the heart of the matter.

The mind is more of a challenge. What we need is a way of tracking impacts that is as effortless and painless as possible. Different people will look for different things. For me, I want to be able to track my impacts on the go, inputting things to my smart phone, whether or not I've got an internet connection, without having to learn a new web interface or having to remember a new username and password. Ideally, I want to be able to record things as I stumble across them, online or in my inbox, without leaving my internet browser or email programme. Each researcher needs to create their own system that will meet their needs, so they can effortlessly keep track of impacts on a day-to-day basis. Not everything they record will necessarily be worth reporting, but when asked to report impacts, they will have a wealth of material to sift through, and be much more likely to provide detailed and comprehensive information. This is not about replacing institutional repositories. It is about finding ways to collate material easily as you go, so it is easier to input to your institutional repository (or whatever form you are asked to report impact in) when that time comes round.

So what are the options? A lot of my colleagues just use their email, putting things related to impact in a folder, filing leads, and emailing notes to themselves to store for safe-keeping. One person I met prints out anything related to his impact and puts it in a ring binder. Others I know are exploring OneDrive. There is no single right answer. We encourage researchers to come up with their own solution so that impact reporting becomes quicker and it is easier to provide high-quality information.

I have developed a system for researchers to keep track of their impacts on a day-to-day basis, prior to submitting them to funders and institutional repositories, using the productivity app, Evernote. Using this system, I collect evidence on the go in three simple steps:

1. Sign up for a paid Evernote account (£30 a year at the time of writing) — only one member needs to do this, the rest of your team can use the free version of the app or website, or just email impacts into your Evernote account.
2. Start a new notebook, share the notebook with your team if they want to record impacts directly into the notebook in their own Evernote app (and anyone else who would like to have access to your impacts e.g. an administrator who is helping you input evidence to an institutional repository).
3. Give your team your unique Evernote email address to send in notes, photos, recordings, documents, clipped web pages and other evidence of impact to be collated in your shared notebook.

Now I have relevant material quickly to hand when I need to report it to funders or my university. My team members don't need to remember a new log-in or learn any new skills; if they can send an email, they can keep track of their impacts (no excuses!). They don't have to download the app, visit a website or even be online unless they want to. For my most recent project, I've made a link on the project website that brings up an email addressed to my Evernote address automatically with a subject line that will deliver their email directly to the notebook for that project (I'm using Evernote to track impact for multiple projects). Evernote is GDPR compliant but as a US-based cloud computing service I am not covered by EU data protection legislation if there is a data breach, so I make sure I don't store personal or confidential data about research participants or anything I have contractually agreed to keep confidential.

We need simple, quick and easy ways of tracking impacts on a day-to-day basis that fit with our busy, often mobile lives (the 'mind' part). We also need to appeal to the heart of the impact agenda — creating benefits for others — and consider how evidencing impact on a regular basis can actually lead to bigger and better impacts. Impact tracking needs to come from the heart *and* the mind if it is to happen regularly and effectively.

If you want to find out more about my impact tracking system for Evernote, visit www.fasttrackimpact.com/evernote.

If you start collecting evidence of impact as you go, evaluating impact will become a lot easier. In the next section, I want to explain how to design an impact evaluation.

What is research impact evaluation?

Research impact evaluation refers to the process of analysing, monitoring and managing the intended and unintended consequences, both positive and negative, of research. Evaluation typically seeks to identify causal links between:

1. The generation of new knowledge through research (or its co-generation with publics or stakeholders);
2. Knowledge exchange activities (via passive dissemination or public/stakeholder engagement); and
3. Impacts, including indirect and unforeseen benefits as well as negative outcomes.

Evaluation may provide direct, sole attribution of impact to research, but more often than not, attribution is indirect and/or partial. As a result, the goal of most evaluations of research impact is to assess the extent to which research made a significant contribution towards an impact.

In this chapter, I distinguish between evaluating and evidencing impact, although the two activities typically go hand-in-hand. Evaluation is the process of assessing the significance and reach of impacts and the extent to which they are caused by research. Evaluation findings that are independent, robust and available for the public to scrutinise can be used as evidence to demonstrate the benefits to society arising from research.

Why evaluate or evidence impact?

Many researchers are content to engage with the public and stakeholders without asking whether the time they spent actually made a difference. They engage with publics and stakeholders because it is the right thing to do, not to get credit for their work, and so there is no need to evaluate or evidence whether or not they helped. But what if it turns out that they made no impact at all and were wasting their time? What if, in fact, they made things worse? If they knew their efforts were failing, they might be able to learn from their mistakes and do better work in future. They might even be able to help fix some of the problems they inadvertently exacerbated.

Whether or not we want or need to report the impact of our research, evaluating our impact can help us engage better with publics and stakeholders, and generate impacts that have greater meaning and value. Evaluation enables us to better understand the interests and priorities of different publics and stakeholders, so we can better meet their needs and provide them with opportunities that they find meaningful, enriching and valuable.

Knowing what works (and what doesn't) can help us choose methods and activities that will engage people more effectively with our research, wasting less time and generating more impacts from the time we do invest. Evaluating our approach to impact can help us anticipate challenges and avoid using methods that are unlikely to work or that might lead to unintended negative consequences. When things don't go according to plan, our evaluation can give us ideas about how to get things back on track or do things better the next time. Whether to funders, the media or our friends and family, evaluating our impact can enable us to communicate the value of research to wider audiences.

What should I evaluate?

Research impacts are typically evaluated against two key criteria: significance and reach. According to the Higher Education Funding Council for England:

- **Significance** of your impact is the extent to which the research has "enriched, influenced, informed or changed policies, practices, products, opportunities or perceptions of individuals, communities or organisations"; and
- **Reach** is "the extent and diversity of the communities, environments, individuals, organisations or any other beneficiaries that may have been impacted by the research".

Note that reach is not just considered in terms of numbers of people reached or geographical reach, but can be considered in more nuanced forms, such as the diversity of organisations benefiting.

Significance and reach need to be framed and argued for, as well as evidenced. What might appear to be an insignificant impact with limited reach may be argued to be highly significant and far-reaching in a given context. If you are able to argue that there is a sub-national need that is unique to and clearly evidenced at that scale, and you fully addressed that need at the scale of the relevant region, you may be able to argue that your impact was significant and far-reaching. On the other hand, if you were to frame the same impact as an international problem affecting every country in the world, but you only solved it for that one region, you may well undermine your argument for a significant and far-reaching impact. I will revisit considerations around narrative and framing later in the chapter in relation to communicating evaluation findings as case studies.

In addition to assessing the significance and reach of your impact, for an evaluation to provide formative feedback to enhance your practice, it is also useful to evaluate the process you follow to reach impacts:

- **Evaluate the design of your pathway to impact:** it is often possible to identify a flawed pathway to impact in advance, if you stop and reflect on the design of your pathway. I once designed a flawed pathway to impact, which included a smartphone app, but with no marketing budget (the app sank without trace among the thousands of apps uploaded to app stores every day). With hindsight, the flaw in my plan is obvious, and I can't help but wonder if I might have spotted this had I taken this step, and evaluated my design more rigorously at the outset. One way to do this is to consider the extent to which: i) the design follows known good practice principles; ii) it is adapted to your particular context; and iii) it is underpinned by sound ethics. For example, a well-designed public engagement process should typically:
  - Identify publics and stakeholders systematically
  - Understand and manage the expectations of these groups
  - Deliver tangible benefits that will be valued by each group in ways that are sensitive to their social and cultural context
  - Identify risks and assumptions and be prepared to adapt to changing circumstances
  - Engage experienced personnel who can manage events, facilitate workshops and organise engagement effectively

- **Evaluate the delivery of activities** along your pathway to impact, and their immediate outcomes. Refer back to the activity indicators you identified in your impact planning template (Step 2, Chapter 10) and choose appropriate methods to track each indicator. This should quickly tell you if you are getting the outcomes you expect, and if not, you will have time to correct your course, and stay on track for impact.
- **Evaluate the impacts of your research:** evaluations that focus only on the delivery of activities along a pathway to impact (e.g. communication reach) and the immediate outputs of engagement (e.g. evaluation of an event) often fail to articulate the broader, deeper and longer-term benefits of research. Typically, this task focuses on evaluating the significance and reach of the impact.

How to do an impact evaluation

To conduct an evaluation of your impact, you need to:

1. Know what impacts you are looking for
2. Select an evaluation design to establish the significance of the impact
3. Determine the reach of the impact
4. Communicate the findings of your evaluation as evidence of impact

The order of these steps is important. Attempts to communicate the impact of research that are not underpinned by rigorous evaluation may unravel under closer scrutiny. An evaluation may be designed to assess whether a project met its original impact goals and miss much more significant impacts that arose opportunistically during the research process. An impact that is shown to have global reach that isn't significantly valued by anyone could be argued to not really be an impact. Instead, be clear on what it is you are looking for, and design your evaluation to determine whether or not there were significant impacts that you can clearly establish were linked to the research. Only at that stage is there any point in investigating the reach and communicating your findings as impact.

These may have been identified at the start of the research, or may have arisen more opportunistically during the research process. Either way, clearly articulate the impacts you want to evaluate. You may wish to frame this as a testable goal (e.g. the research made a significant contribution towards impact), question (e.g. to what extent did the research contribute towards impact?) or null hypothesis (e.g. that the research made no discernible contribution to impact). To ensure you have holistically identified all relevant impacts, you may want to revisit the list of different impact types in Chapter 2 to consider if there are any missing types of impact you might want to evaluate.

2. Select an evaluation design to establish the significance of the impact

The next step is to determine whether or not (or to what extent) the research contributed towards significant effects or impacts. It is not enough to demonstrate that impacts occurred. It is essential to be able to prove beyond reasonable doubt that the impacts came about as a result of the research. To do this, evaluations typically seek to identify causal links between the generation of new knowledge through research (or its co-generation with publics or stakeholders), knowledge exchange activities (via passive dissemination or public/stakeholder engagement) and impacts, including indirect and unforeseen benefits as well as negative outcomes.

To do this, guidance from the realms of evidence-based policy and research-informed international development typically follows a hierarchy of methods, based implicitly on their accuracy and lack of bias. Randomised controlled trials sit at the top of this hierarchy, followed by quasi-experiments, mixed methods and qualitative methods. Implicit in this hierarchy is the idea that quantitative measures are superior to qualitative approaches, and the task of evaluation is to identify and evidence the sole cause of any given effect, where the cause is an intervention based on research and the effect is the impact.

However, it is increasingly clear that the relationship between research and impact is far more indirect, non-linear and complex than these evaluation frameworks allow. Demonstrating cause and effect can be tricky in the real world. There are always many other factors that may have been responsible for the impacts you would like to be able to claim as your own. Many of the benefits that accrue from research take years to materialise. Impacts may become evident long after project funding has ended, making it difficult to find the staff time or funding to evaluate impact properly. Some impacts appear self-evident and can be evidenced with data collected and published by others. Other impacts are difficult to define or measure, and require a research project of their own to evidence credibly.

I have therefore identified eight different types of evaluation design that can be used to establish cause and effect in different ways (Table 6). Broadly speaking, research impact evaluation methods can be characterised along three continua:

- A summative focus on evidencing and claiming impacts and being accountable (sometimes referred to as external evaluation), versus a formative focus on learning, adaptation and taking epistemic responsibility for the generation of impact (internal evaluation)
- Sole attribution versus significant contribution: tracing pathways to impact and assessing the significance of the contribution
- *Ex-ante* efficacy assessments of what in theory will work, including anticipation of the impacts to be generated, versus *ex-post* effectiveness assessment of what works in practice

Each type of impact evaluation in Table 6 is located in a different place along the three continua described above. They each take a different approach to establishing cause and effect (between research and impact), and give rise to different forms of evidence. Many of the methods can be used to monitor impacts as they arise as well as evaluating impacts after the event.

Select an evaluation design based on the sorts of data you think you will be able to obtain, and the types of impact you want to evaluate. It is possible to use more than one evaluation design in a mixed methods approach to impact evaluation, e.g. case-based evaluations typically integrate a number of methods as part of an overarching narrative (more on this approach later in the chapter).

Evaluation may provide direct, sole attribution of impact to research (e.g. via an experimental or statistical evaluation design). However, more often than not, attribution is indirect and/or partial, requiring more nuanced approaches to evaluation, such as contribution and pathway analysis, evidence synthesis, or participatory and arts-based methods (Table 6). As a result, the goal of most evaluations of research impact is to assess the extent to which research made a *significant contribution* towards an impact.

Justifying the significance of the contribution may be done via statistical inference of a proportion of impacts that can be attributed to the research (e.g. impacts above a baseline after publication of the research). Depending on the type of impact you have generated, this sort of evaluation can be particularly useful for assessing indirect impacts (e.g. where the research stimulates other activities that ultimately lead to the impact) and cumulative impacts (e.g. where the impact of the research is dependent upon other impacts that occur in parallel based on other sources of evidence).

However, it is often more feasible and appropriate to build a more multi-faceted argument about the extent to which the contribution can be argued to have particular value or meaning in a specific context (e.g. closing a small legal loophole that was costing taxpayers millions or contributing towards the design of a specific policy mechanism within a new directive that led to substantial benefits), supported by multiple forms of evidence including testimonials. For example, for an indirect impact, a researcher may be able to trace a pathway from the impact to their research via multiple causal links in a chain of events all the way back to the research. If the first causal link from the research to the rest of the chain can be demonstrated to be directly linked to the research and each of the subsequent links can be shown to be dependent on that first causal link, then a robust argument can be made that the research made a significant contribution towards the ultimate impacts.

I have provided examples of methods you may want to research and try out for each of the eight different types of evaluation. These include a mix of qualitative and quantitative methods. There is no single right approach, and many researchers adapt methods from their own disciplinary toolkit to evaluate impact, or stick with types of method (e.g. qualitative or quantitative) that they feel most comfortable with. In addition to feeling comfortable with the methods you select to as part of your evaluation design, it is important that they are suitable for assessing the types of impact you are evaluating. Play 'devil's advocate' and ask yourself what you would need to do to convince someone who does not believe that your work has led to any sort of impact. In some cases, to be convincing you will need quantitative evidence, for example, an increase in visitor numbers and museum revenues after the installation of a new exhibit based on your research. In other cases, qualitative evidence will be more convincing and appropriate, for example, illustrative quotes describing how engagement transformed people's attitudes towards an issue or group of people.

3. Choose methods for determining the reach of the impact

The reach of an impact typically extends in two ways:

- Impacts may 'scale out' if they spread from one individual or community to another, for example, as people pass evidence to colleagues or adopt a new research-based innovation.
- Scaling-up happens when an impact reaches a higher institutional or governance level (e.g. from a delivery agency to a government department), or a wider spatial scale (e.g. widening the reach of an impact from a farm to catchment level).

Evaluation methods need to be adapted to the type of scaling process through which reach occurs. For example, I recently assessed institutional scaling-out via a combination of quantitative Social Network Analysis of research findings as they were passed from person to person through policy and practice networks, combined with qualitative interviews to understand what was passed to whom, how and why. For another impact evaluation, I wanted to understand how my research was scaling up from local pilot projects to a national scheme, and so set up a registry that all new initiatives based on my research had to join, to track the impacts they achieved. For many impacts that scale up geographically, there will be third parties collecting data (e.g. public statistics) that can be used to infer reach. Sometimes you have to collect this data yourself. For example, I designed an impact evaluation for a colleague who wanted to be able to demonstrate that his research on the health benefits of organic food was influencing purchasing decisions across Europe, and we had to commission a large-scale survey of consumers in two countries before and after publication of his work to be able to infer cause and effect, before then being able to use European data that showed a spike in the consumption of organic products after the publication of his research. In Part 4, I have described how you can design an evaluation to find out how your work might be being used by policy-makers around the world.

4. Collect and analyse evaluation data

You don't have to wait till the end of your research to start collecting evaluation data. While you're still on your pathway to impact, engaging with publics and stakeholders, start collecting data to look for specific planned impact milestones. Build opportunities for longitudinal evaluation into your work, for example, incentivising participants to provide you with their email addresses (e.g. via a prize draw or joining a mailing list that provides additional free benefits or opportunities), so you have the opportunity to re-engage people to deepen and broaden your impact, and follow-up with surveys or interviews later to find out if longer-term impacts have arisen from your work.

There are often important opportunities for formative feedback if you evaluate impacts during the research cycle (see 'what should I evaluate' above). The types of evaluation and methods that are identified in Table 6 are designed for assessing progress towards pre-identified impacts. You can expand the range of impacts you are evaluating, if you think the original impact goals were too narrow. However, it is important to also collect data opportunistically as impacts arise that you are not expecting. To do this you will need some sort of system for storing material you think might be relevant later, quickly and easily (see the next section).

Finally, it is important to keep some perspective and bear in mind that not all impacts require the same level of evidence. For highly controversial, multi-factorial, contested or high-profile impacts, such as a new drug discovery or genetically modified crop that increases yields whilst reducing pesticide and fertiliser use, you might be expected to supply a heavy burden of proof. For more obvious impacts where there is a clear theory of change leading to an apparent impact, a lighter burden of proof may be appropriate. For example, if providing mains water to a village reduces the amount of time households spend collecting water, the only plausible explanation is the improved proximity of water. Similarly, the introduction of hand-washing in hospitals was based on a dramatic reduction in maternal mortality observed in a hospital in Vienna in the 1840s, despite the fact that the germ theory of disease had not yet been proposed and so there was no way of proving cause and effect. In some cases, simply being able to triangulate more than one source of evidence, even if that is only based on a strong testimonial, may be all that is necessary to attribute impact to research convincingly. If you are not sure whether you have done enough to demonstrate your impact, get a second and third opinion, and test how well your arguments stand up to scrutiny.

Table 6: Types of research impact evaluation with examples of commonly used methods, defining characteristics, approach to establishing cause and effect (between research and non-academic impact) and examples of the sorts of evidence they give rise to

Type of research impact evaluation	Examples of commonly used frameworks and methods	Characteristics	Approach to establishing cause and effect	Examples of evidence	Types of impact typically evaluated
Theory-/logic-driven	Theory of Change; Logical Framework Analysis; Fast Track Impact Planning Template (Table 3, Chapter 10) and other logic models	Can be used in summative or formative mode, typically *ex ante* (but can be used *ex-post*), to show contribution rather than sole attribution	Generative causation, identifying causal processes in chains from the generation of research to impacts in the context of wider supporting or mediating factors and contexts	• Indicators will be set at the outset to monitor progress along proposed pathways to impact, e.g. hectares of land restored, followed by reduction in water pollution, savings to water companies and reductions in water bills	• All types of impact
Experimental and	Randomised controlled trials; quasi and natural experiments; regression	Typically used in summative mode, *ex ante*	Counterfactual causation based on the difference	• Improvements in water quality based on	• Economic impacts

quantitative survey work	discontinuity studies; monetary and non-monetary (deliberative and non-deliberative) choice experiments; non-monetary valuations (e.g. multi-criteria evaluation); economic (game) experiment; other forms of social and economic impact assessment; direct measurement of key variables pertaining to impact via sampling of treatments versus control; indirect measurement via indicators to infer impact; model-based approaches, calibrated and validated with more limited sampling; questionnaires (e.g. evaluating levels of awareness before and after engagement such as entry/exit quizzes or before and after Likert scale questions to quantify the extent to which attitudes	and/or ex-post, to infer sole attribution or quantify the extent to which an impact can be attributed to research	between two otherwise identical cases (including individuals, sites, environments/contexts), one that is manipulated and the other that is controlled giving rise to evidence of cause and effect	improved regulation arising from research • Reduced morbidity and mortality among patients receiving new treatment based on research compared to control group • Cost savings attributed to a new treatment or intervention compared to a baseline or control • Changes in awareness or attitudes before and after engagement • Monetary benefits arisen from a change on land manage practice	• Environmental impacts • Health and well-being impacts • Policy impacts • Other forms of decision-making and behaviour change impacts

	Tools/techniques	Use	Description		Impacts
	have changed), website or in-app questions based on interactions with website or app content, multiple-choice questions (e.g. before, during and/or after public lectures); photo survey technique			informed by research • Optimization in the choice of policy instrument to promote a public health intervention, informed by research	
Statistical	Statistical modelling; longitudinal analysis; econometrics; Payback Framework; difference-in-difference method; double difference method; propensity score matching; analysis of distributional effects; value of information analysis; Cost-Benefit Analysis; value for money measures; analysis of secondary data showing evidence of behaviour change (for example,	Typically used in summative mode, a piori and ex-ante, to infer sole attribution or quantify the extent to which an impact can be attributed to research	Causation inferred from correlation between cause (dependent variables) and effect (independent variables) or statistical difference between effect before/after or with/without an intervention (cause), controlling where possible for confounding effects,	• Numbers of companies, employment or new roles in the workforce • Numbers of (or profits from) new commercial products or spin-out companies • Improvements in indicators of social cohesion or social mobility	• Economic impacts • Environmental impacts • Health and well-being impacts • Policy impacts • Other forms of decision-making and behaviour change impacts

	purchasing behaviours); analysis of secondary data showing changes in capacity (such as access to resources that were not previously available)		and quantifying the extent to which effects can be attributed to multiple causes	• Probability of time, money, species, or lives saved as a result of new evidence-based practices
				• Policy impacts • Other forms of decision-making and behaviour change impacts • Capacity or preparedness
Contribution and pathway analysis	Contribution analysis; knowledge mapping; Social Network Analysis (e.g. before and after your work, showing larger, more connected networks); Bayesian networks; agent-based models, Dynamic System Models; influence diagrams; changes in policy linked to your research (evidenced via mentions and citations in policy documents and/or testimonials from policy-makers)	Typically used in formative mode to provide feedback to enhance impacts in progress, as a way of assessing the extent to which research contributes to impact	Additive causation based on tracing links between causes and effects along causal chains or pathways to impact	• A significant contribution made by research to the solution of a previously intractable problem • A significant contribution to the development of a new policy mechanism • Increase and strengthening of the number of nodes or connections in a social network

Case-based and narrative analysis	Case studies; testimonials; ethnography; participant observation; qualitative comparative analysis; linkage and exchange model; content analysis or Grounded Theory Analysis of interviews, focus groups and vox pops; Q methodology to quantify and categorise people who identify with contrasting attitudinal statements derived from interviews; opinion polls; qualitative analysis of written comments (e.g. left in visitor books, feedback forms, written on a graffiti wall or written on postcards that are posted back to participants after the event)	Typically used in summative mode, ex-ante, to assess the extent to which research contributes to impact	Causation is inferred by building a case that triangulates multiple sources of evidence (potentially including the outcomes of other types of evaluation, such as statistical or experimental) to create a credible argument for a significant contribution of the cause to an effect, sometimes based on the narrative of a theory of change	following a participatory process	• Citations relating to policy-making processes, new legislation or a change in the law influenced by the research • Evidence that policies have been implemented and are meeting their objectives • Improved compliance with regulation • Testimonials from practitioners explaining how they gained capacity (e.g. improved skills,	• All impact types

			understanding and confidence levels) that enabled them to enhance their practice	• Environmental impacts • Health and well-being impacts • Policy impacts • Other forms of decision-making and behaviour change impacts • Cultural impacts • Capacity or preparedness
Participatory	Participatory monitoring and evaluation; empowerment evaluation; engagement evaluation; action research and associated methods	Typically used in formative mode to enable beneficiaries to engage and shape feedback that then enhances impact, while impact generation or ex-ante is in progress, as a way of assessing the extent to which research contributes to impact	Causation is inferred by jointly building a case with beneficiaries that triangulates multiple sources of evidence (including data collected by beneficiaries) to create a credible argument for a significant contribution of the cause to an effect	• Improvements in variables that indicate the achievement of goals set by a stakeholder or other social group who co-produced research (e.g. number of community members having acquired a particular skill) • Changes in perception of a resource management issue

Evidence synthesis	Meta-analysis; narrative synthesis; realist-based synthesis; rapid evidence synthesis; systematic reviews	Used in summative mode, ex-post, to infer sole attribution or quantify the extent to which an impact can be attributed to research	Cumulative causation based on the systematic aggregation and analysis of cause and effect across multiple evaluations (of any type) in different contexts	• Time, money or lives saved as a result of new evidence-based interventions or practices • Evidence that an intervention works across multiple contexts	• Economic impacts • Environmental impacts • Health and well-being impacts • Policy impacts • Other forms of decision-making and behaviour change impacts
Arts-based	Ethics (philosophy); linguistics; aesthetics; creative expression; oral history; story-telling; digital cultural mapping; (social) media analysis	Can be used in summative or formative mode, while impact generation is in progress and/or ex-ante, to assess the extent to which	Causation is inferred by building a case that triangulates multiple sources of evidence (potentially including the outcomes of other types of evaluation, such as statistical or	• Changes in awareness or attitudes of a social group as a result of engaging with research • Changes in culture, cultural discourse or	• Environmental impacts • Health and well-being impacts • Policy impacts • Other forms of decision-

	research contributes to impact	experimental) to create a credible argument for a significant contribution of the cause to an effect	appreciation and benefit from cultural artefacts and experiences	making and behaviour change impacts • Cultural impacts • Capacity or preparedness

Communicating your evaluation as evidence of impact

In some cases, you will be able to use your evaluation findings directly to evidence your impact, for example if you conducted a randomised control trial, or if you are able to submit raw data as evidence. However, it is often necessary to take an additional step to analyse and publish evaluation findings in a way that makes them publicly available, and which you can then cite as evidence to support impact claims. You may have conducted an impact evaluation survey, and as a result, you may now know that your research has had an impact. However, it may not be reasonable to expect others to take this on trust, if the evidence is a pile of completed questionnaires sitting on your desk. By analysing and publishing your findings, you turn your evaluation into evidence.

Depending on how controversial or important your findings are, to be believable, you may need to consider how you publish them. For example, publishing your evaluation findings as a blog or on your own website may be an acceptable way of opening your evaluation findings to public scrutiny for a small project that is not making particularly controversial or significant claims to impact. However, for a large project that is making controversial or significant impact claims, it is not unreasonable to expect a more detailed report to be published more formally. For example, you might publish your findings as a report co-branded by your institution and project partners, an independent report written by a consultant and published by your project partners, or as a peer-reviewed article. I have written in greater length about these options in Part 4, "Writing up an impact evaluation as a research article".

A case study is an effective way of communicating the wide range of impacts possible from research, using a diversity of evaluation methods. The world's largest database of research impact case studies was published in 2014, containing over 7000 cases (http://impact.ref.ac.uk). For UK readers, I have provided a guide to writing a top-scoring case study for the Research Excellence Framework in my guide in Part 4 of this book. In summary, based on my analysis of high- and low-scoring case studies from the 2014 database, impacts in top-scoring case studies were:

- Significant;
- Far-reaching;

- Clearly articulated;
- Convincingly evidenced; and
- Focused on the benefits rather than the pathways to impact.

Creating an impact case study is partly about having high-quality evidence to corroborate your claims of significance and reach. However, it is also partly about the narrative you create, as the following examples illustrate.

'Turner's Yorkshire' is an example of impact arising from research in fine art. Professor David Hill from the University of Leeds published extensively on Turner's work, highlighting Yorkshire as a landscape of international significance. His fieldwork tracked the artist's travels through the county, locating, examining and photographing his viewpoints as they survive today. A tourist promotion, 'Discover Turner's Yorkshire', gave this work much wider public impact, with published and online materials, such as the Turner Trails website (with walking routes and audio guides) raising public awareness of the significance of the county to the artist (Figure 20). This increased tourism and brought economic and social benefits, which the researchers quantified as far as possible in their case study.

Figure 20: Screenshot from the Turner Trails website

What interests me about this as an example of an impact case study is the comprehensive and innovative use of evidence:

- 100,000 page views, 10,000 downloads
- Estimated 1.25 million visitors saw interpretation boards
- Visitors to Turner Trails spent on average £199 per head per trip
- Over 50% of local tourism businesses thought the project had a positive effect on business
- Extensive media coverage equating to £600,000 in total Advertising Value Equivalency (note: this measure is not viewed as being credible nowadays)

Cardiff University's DECIPHer-Assist project claims to be the UK's most effective school-based smoking prevention programme. Peer-nominated students aged 12–13 were taught how to intervene as 'peer supporters' with their Year 8 peers in everyday situations to discourage them from smoking. The impact of this education research was given the highest possible grade in the UK's Research Excellence Framework. Evidence of the impact included:

- Over 60,000 students have taken part since 2010
- Cited as good practice in policy documents
- Cardiff research suggests 1,650 young people will not go on to take up smoking as a result
- Treatment of lung cancer in England costs £261M per year. If implemented throughout the UK, DECIPHer-Assist would prevent 20,000 young people taking up smoking each year
- Award-winning company set up to licence the programme

A University of York sociology project called 'Advising the advisers' helped improve the conduct of adviser-claimant interviews in Jobcentres. This impact was also awarded the highest possible grade for its significance and reach. Policy-makers learned about evidence via working papers and presentations, and changes in policy resulted from the work, including new procedures and compulsory training for advisers. This impact was evidenced using testimonials from those who had benefited from the work, such as this one, from a senior civil servant:

"This research has had impacts in immediate and potentially long-term performance gains. We are now using the results of this

research to develop and test [an evidence-based] adviser training programme. The results of this research have the potential to change the whole adviser training approach"

These examples show the wide range of different types of impacts and evidence that can be used in case studies. They also show how differently impact is evidenced in different disciplines. You can read thousands more in detail at: http://impact.ref.ac.uk. It is worth dipping into this database. Browse through case studies in your subject area or search for keywords you are working on if you want some inspiration. Read through the descriptions of impact to get new ideas about types of evidence you could collect and use to communicate your own impacts. If you are interested in how to write a high-scoring case study for the next Research Excellence Framework, I have written a guide in Part 4.

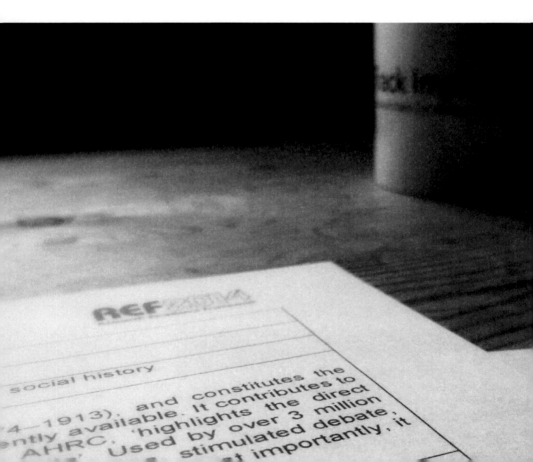

Chapter 23
Conclusion: Left hanging (with the right tools for the job)

One of the most powerful metaphors for my approach to impact is this image of a man who has been lowered over the edge of a roof to fix a sign (Figure 21). You have to assume that these two men know and trust each other, given the level of trust that the dangling man has clearly placed in his colleague. One can imagine the conversation that preceded this operation, as they discussed what might be wrong with the sign, and who would volunteer to be lowered into position.

Figure 21: Two men fixing a sign on a Russian furniture store

For me, this illustrates the heart of this book: impact is based on long-term, trusting, equal relationships, and two-way communication between researchers and those interested in our work. As we work together, we negotiate our goals and carefully plan how we will reach them, considering the roles we will each play in the process.

Only in the context of such long-term, two-way and trusting relationships does it become relevant to ask if we've got the right tools for the job. In the case of the two workmen, we can assume that the man dangling over the roof knew that he would need a screwdriver. We might also assume that if he's holding a flat-headed screwdriver in his hand, he might have a different type of screwdriver in his pocket, just in case. When we're planning for impact we need the right tools for the job, and it is always good to have a plan B, in case the activity we planned doesn't deliver the desired results.

I want to make an important point of emphasis with this metaphor. It is easy to skip to this third section of the book and pick and choose tools and techniques to achieve impact. But if we do this, there is a real danger that we miss the whole point of impact, which is to create social and economic benefits that make the world a better place. Impact then becomes a box-ticking exercise in which stakeholders and members of the public are left feeling used and bemused by their interactions with us. There are plenty of toolboxes available that will inspire you to try all manner of exciting new knowledge exchange activities. But if our focus is on the toolbox rather than the context in which to use those tools, then we may be tempted to drop the people we are working with when the project ends. Like the man holding his colleague for as long as it takes, holding onto relationships for the long term can sometimes be hard work. But if you invest in relationships for the duration, you'll be around when the opportunity arises for your research to be put into action. And the people and organisations you're working with will be around when you need a letter of support for your next grant application.

Research impact is a collaborative endeavour — we can't do it on our own. So, let's swallow our pride and adopt an attitude of service that identifies and responds to opportunities as they arise. If researchers start co-producing knowledge with the people who need it most, the ideas we come up with really will change the world. None of us has to do anything particularly big, but taken together, we have enough collective intelligence to solve almost any problem, if we will just work together.

You don't have to be a natural communicator or extrovert. You don't even need to be that confident. Your research can make an impact, even if you are chronically shy, if you have the heart for it, a plan and a few relevant skills. My 'eureka' moment came at a time in my life when I was facing the causes of my own chronic lack of confidence. Having experienced sexual abuse throughout my childhood, I had very little sense of 'self', let alone self-confidence. As an academic I lived in constant fear of what others thought of me. Each professional failure appeared to confirm the sense of self-loathing that was always lurking beneath the surface, no matter how bright and cheerful a face I put on. I found engaging with stakeholders terrifying for years. I would sit outside people's offices with sweaty palms and heart thumping, feeling sick before entering each meeting. The fear only became greater with time, as the people became more important and the stakes higher. But I pushed through that fear, believing that through my actions I could give many times more to the world than that one person had taken from me as I grew up.

All of us have something we can give back to the world, no matter how small and insignificant we might feel our research is. No matter how much fear the steps in this book might hold for you, it is possible to push through this and take those steps towards impact. You don't need to change who you are, or try and become like anyone else. You can take these steps, whether you are a scientist or an arts and humanities scholar; a PhD student or a professor; confident or terrified. Unfortunately for me, being me has meant embracing the fear. However, it turns out that 'just being me' makes people more likely to take my research seriously and engage with me. That's one of the reasons that I wanted to write this handbook very much 'as me' rather than as a dry, third-person account of how to generate impact from research. I can't put myself in the shoes of everyone reading this book, but it is my hope that you are now eager and ready to put yourself in the shoes of those who might use your research. If you feel overwhelmed by the amount of information in this handbook, remember that all you have to do is connect with the people who are interested in your research and the rest will come naturally.

The steps in this book, underpinned by each of the five principles, have the power to enable all of us to start generating and sharing knowledge in new ways. As we work more closely with those who are interested in our work, or who want to use it, we will learn how to generate knowledge that is more interesting and useful, and can effect change in ways we would never have imagined possible. Even if each of us only makes a small change through our research, if we can be that change, together we have the potential to change the world.

To me, this book is a bit like a dandelion. To some it will be an inspirational wildflower, while to others it will be a weed for the compost heap. But for me it is a wish, as I blow its seeds into the wind and hope that a few of them take root in your research. It is impossible to predict which way the wind will blow these seeds, but that's what makes sharing ideas so exciting.

simply disseminating information, we need to actively engage with those who are looking for new insights, understand their needs and work with them to co-produce new ideas that can actually shed light on real-world issues. Some of the most beautiful candles can be instantly extinguished by a gust of common sense when we take them out into the real world. But if we take this chance, we might just hit on an idea that catches on like wildfire.

A relational approach to impact

My point is that generating new knowledge isn't enough; we have to learn how to share our knowledge if we want it to be used and to generate impacts. Making our research available online isn't enough. People need to *learn* about our research if the data and information we produce is to turn into knowledge that can be applied in the real world. To do this, we need to patiently nurture relationships with those who are interested in and can use our research. This takes humility, because we need to listen and learn if we want to understand who these people are, and what motivates them. Having letters before or after our name does not makes us any better (or worse) than anyone else. We have probably all been in situations where people have given us undue respect as a researcher, and in others where we are instantly mistrusted because we are researchers. If you can be yourself, you can usually cut through such preconceptions in time. It is from that place of equality that you can build the kind of two-way, long-term, trusting relationships that can enable people to learn about and apply your research.

This approach stands in stark contrast to the concept of 'knowledge transfer'. Knowledge transfer treats new knowledge like a 'gift' that can be transmitted unchanged from one person to another. If knowledge is information that people have learned and know about, then people may interpret information in different ways as they learn about it. Knowledge changes as it passes from person to person through social networks, as people adapt it to their own contexts and needs (or in some cases cherry-pick and twist it for their own ends). Even if it were possible to pass the gift of knowledge on unchanged, this approach assumes that the person receiving the gift will appreciate it and be able to use it, despite the giver knowing little or nothing about their needs and preferences. We all know what it is like to be on the receiving end of well-intentioned but ill-informed gifts, which we know we'll never use.

Although there are some situations where it is appropriate to simply communicate research findings to the people who might use them, there are very few situations where some level of dialogue with these people wouldn't improve the flow of knowledge. If nothing else, talking to those who are likely to use the knowledge you are generating can improve how you communicate with them, and enable you to better tailor and target information. The reality is that you will probably want to engage with people who are interested in your research in different ways at different points in the research cycle. For example, you may shape the initial research through intensive dialogue with a few key players, giving way to more extensive communication towards the end of a project, so that your findings reach as wide an audience as possible.

We now live in a digital age where the rules of public and stakeholder engagement are being rapidly rewritten. Understanding the power of these new tools and how to use them cleverly and responsibly can enable you to engage with groups that would have previously been inaccessible. This isn't about replacing face-to-face contact. If you want to be truly influential as a researcher, the warmth of a handshake and a shared conversation over coffee or in the corridor is just as important as it has always been. But by combining the efficient use of new media with everything we already know about working effectively with stakeholders and the public, we can do much, much more than ever before. Furthermore, contrary to popular belief (at least amongst the researchers I typically train), it is possible to spend *less* not more time at work when you engage with these new approaches to work and communication (see Chapters 17 and 18). Working with people to generate impacts from your research *will* take time and patience. However, the very tools we can harness to help us generate those impacts can actually save us time elsewhere in our day. That means we get time to engage with the outside world without wrecking our work-life balance. You will be surprised at how fast you can achieve some impacts from your research, if you try out some of the things I'm going to suggest in this book.

Evidence-based principles to underpin and fast track your impact

This book takes an evidence-based, relational approach that enables you to fast track your impact whatever your career stage or discipline. Based on research with researchers, knowledge brokers and stakeholders in different research contexts around the world, my colleagues and I have distilled five principles to underpin impact in the first section of this book. In the second section of the book, I have linked these principles to five practical steps, so you can deliver significant, far-reaching and lasting impacts from your research (Figure 1). Here, I want to outline each principle and explain how it can help you generate impact more effectively in ways that save you time.

Principle 1: Design. The first principle is to know the impacts you want to achieve and design impact into your research from the outset. I think most of us are pretty good at coming up with research questions and objectives, but we're not used to setting objectives for our impact. If you have a clear idea of exactly what change you would like to see as a result of your research, you can make a plan to get there, and you're immediately much more likely to achieve impact. This can save you time because it empowers you to look more strategically at impact. You can now see multiple pathways that could take you to the same impact, so you can choose the path that is easiest, fastest or most cost-effective, depending on your key constraint. The first step, which I'll introduce in Chapter 4, will enable you to start thinking critically about the impacts you would like to see as a result of your research.

Principle 2: Represent. The second principle emphasises the value of systematically representing the needs and priorities of those who might be interested in or use your research. Many of us are fairly sure we already know who is most likely to be interested in or might benefit from our research. If not, then we'll typically open our address books and the address books of our colleagues to get some ideas of the sorts of people we might want to engage with. The problem is that most of us only have fairly vague ideas about the sort of people who might be interested in our work outside academia. And we often forget that the address book approach is likely to be highly biased towards certain groups of people and may lead us to overlook important groups who would have been interested if we had only identified them at the start. You may only have time to reach out to one person per month over the next six months, but by doing a publics/stakeholder analysis you will have a ranked list of the groups or organisations you need to reach out to first. You may run out of time after the first six months and only reach out to six groups, but you will know that the 15 minutes per month you spent on this activity were well spent, and you will be significantly closer to impact. Linked to this principle, the second step, in Chapter 5, will show you how to systematically identify 'stakeholders' and identify public audiences for your research. These include the 'beneficiaries' we typically think of first, as well as groups who may be disadvantaged or negatively affected by our research, and those who may have the power to enable us or block us from completing our research and achieving our impacts.

Principle 3: Engage. The third principle is the most important. If you were to boil this whole book down to a single word, it would be 'empathy', and it is encapsulated in this principle. To have an impact, you need to build long-term, two-way, and trusting relationships with those who will use your research, so you can ideally co-generate new knowledge together. This is about having two-way dialogue as equals with the likely users of your research, not lecturing them or doing 'knowledge transfer'. You need to think of ways that you can maintain relationships beyond the typical life-cycle of a PhD project or research project, for example, by engaging colleagues who will be in the post for the long term, and by continuing to engage between projects via social media, newsletters and offering seminars etc. This approach will pay dividends in the end, whether in terms of future jobs and collaborations, or in terms of getting that crucial letter of support for your next research proposal. You can save time by building relationships early on with 'knowledge brokers' who sit between multiple networks of stakeholders or publics that you wouldn't otherwise have access to. They may be able to open doors and short-cut you to trusting relationships based on the recommendation from a trusted intermediary. However, there is no denying that investing in relationships takes time. So the third step, in Chapter 6, will help you to become significantly more efficient so that you have more time to engage in impact-generating activities, whilst also improving your work-life balance.

Principle 4: Early impact. The next principle might seem self-evident, but we need to remember that what we might view as impact might be quite different to the people we want to benefit from our research. In particular, researchers have a habit of thinking of impact over periods of at least three years, whereas many of the people we want to work with will be expecting impacts in weeks and months. Partly this is about managing expectations, but it is also about trying your best to deliver tangible results as soon as possible, which can help keep people engaged with your work. There are a bunch of quick wins that most of us can provide fairly easily, for example, early publication of literature reviews (where possible turning these into more digestible briefing notes) and co-ordinating research milestones to match the time-horizons of decision-makers. I will explore this in more depth in the fourth step, in Chapter 7.

Principle 5: Reflect and sustain. Finally, you need to keep track of what works, so you can improve your knowledge exchange, and continue nurturing relationships and generating impacts in the long term. You can save time by avoiding repeating the mistakes of others when you share what works, as well as your failures, with colleagues. If you have been keeping track of your impacts, it becomes much easier and faster to report impacts when asked by funders. In the second step, you will come up with an impact plan that integrates simple indicators you can use to track whether or not your activities are taking you forwards or backwards along your pathway to impact. The last, the fifth, step in Chapter 8 is about regularly reflecting on your knowledge exchange and impacts with your research team and key stakeholders, so you can learn from your peers and share good practice.

In the third part of the book, I have provided a guide to some of the latest and most important tools and techniques you can use to implement each of these steps. Finally, in Part 4, I have tried to cover all the other questions I am regularly asked about research impact, from eliciting testimonials about your work to turning your paper into an infographic. I have left some topics outside the scope of this book. Most notable are questions around intellectual property, copyright and patenting your research. I will give you the basics that you need to know in Chapter 4, but this is a complex area and there are many good books that can give you far more detailed advice than I have given in this book. I have also not covered the full range of approaches and technologies for communication, including, for example, writing press releases, podcasting and writing for non-specialist audiences. There are many books, websites and courses that can help you in all of these areas.

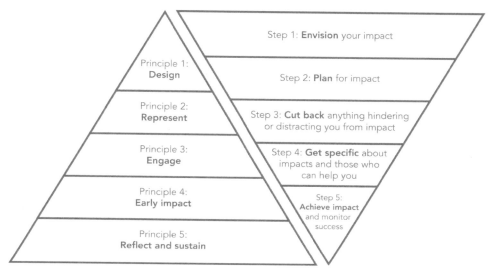

Figure 1: Principles to underpin impact and steps to fast track impact

Start planting your ideas

This isn't a book of theory. It isn't something to bear in mind for the future. It is intended to be used by you today in the research you are designing, conducting or writing up now. So, before you move to the next chapter, choose a research project that you would like to have more of an impact. If possible, choose a project that you're leading or a part of a project over which you have control or influence.

I've written this book because I want to change the way researchers generate and share knowledge, so that their ideas can change the world. I hope this book gives you the tools to enable you to achieve this.

You can plant a seed and it becomes a flower. You can plant an idea and it becomes another's. An idea really can change the world.

Will you be that change?

Part 4: Templates and Examples

Getting testimonials to corroborate the impact of your research

Nobody likes soliciting compliments, so it can feel awkward asking someone you've worked with for a testimonial to corroborate the impact of your research. As a result, most researchers put off requesting this information until they absolutely have to. By this time, the person who was familiar with their work may have moved on, retired or passed away. For this reason, researchers are commonly advised to request testimonials sooner rather than later.

Luckily, there are a number of options open to you when requesting testimonials that can make it easier to ask and more likely that you'll get answers you can actually use:

1.   Don't hide the reason you are asking for a testimonial, but rather than framing it in relation to your career or institution's goals, ask people to reflect on how you have helped them achieve *their* personal or strategic goals, and offer to help them further. Ask for both positive and negative feedback so you can improve your practice and help them better. Instead of this feeling like you are fishing for compliments, you can now enter into a more balanced conversation.

2.   Most people make fairly vague requests using open questions, and are then surprised and disappointed when the answers are vague and lack relevance to the criteria against which they will be judged. Instead, provide additional probing questions, based on the types of impacts you expect might have occurred (e.g. were jobs created, did you save or lose money or become more or less profitable, did attitudes change?), and use these to ask for other specific examples of things that changed or had value or meaning. Ask people to be as specific as possible, quantifying benefits or justifying their significance or reach in other ways. Ask if they know of any evidence you could use to support their claims. Ask them to comment on the specific role your research played, if there were multiple factors influencing the impact. No matter how specifically you probe, always remember to also ask open questions to capture things that you did not expect and so would not think to probe for.

3. Ask for revisions to testimonials that are too vague or not sufficiently relevant. If the person was being vague on purpose because they did not perceive any significant benefit, then you will have to accept this, but people are often happy to provide additional details if asked. It is possible to create specific, probing questions that are not leading questions (see the examples in the previous point), and so not put words in people's mouths. Some researchers draft testimonials for people to adapt, but this may significantly bias the testimonial and limit the spontaneity and breadth of responses. It is far better to probe and request revisions, or if the person does not have time to reply via email, pick up the phone...

4. Interviews are less widely used, but highly valuable for collecting testimonials. They may be done in person, by phone or by Skype, and can be an efficient way for busy people to engage with the process. There are two key benefits to interviewing people for testimonials. First, interviews are a great way to probe for impacts that are difficult to articulate, such as changes in attitude or well-being, or cultural change, enabling the researcher and interviewee to explore the questions creatively to reach quotable conclusions that would not have been offered via the written word. Second, they give the researcher the opportunity to ask open questions to explore the possibility of impacts that they had not planned for or expected. The danger of taking a highly structured approach, asking only for anticipated impacts, is that people do not report the additional benefits you are not looking for (and these may be significant). Open questions during interviews provide the space for people to think about and explore the breadth of impacts arising from your research. Seek informed consent to record the interview and transcribe it, pulling out key quotes and sending these to the person to check and amend as the basis of their testimonial. If possible, ask them to send this back to you on letter-headed paper once they are comfortable with the content. Some universities use video interviews to collect testimonials, which they put online as YouTube videos and link to in the 'corroborating sources' section of their impact case study.

5. Finally, you may want to get a concise, quotable summary of the key points made in the testimony. You can ask for this, or you can propose your own summary based on their testimonial and ask them to amend this so they are happy with it.

Testimonials are normally provided on letter-headed paper, but it is worth noting that 743 case studies in the UK's Research Excellence Framework in 2014 (REF2014) submitted email testimonials. If necessary, auditors may follow up emails with those who sent them to verify their legitimacy, so you had better hope that they still answer to this email address if you want the evidence to withstand scrutiny.

In REF2014 there were three levels of redaction you could apply for, to enable sensitive material to be reviewed, including redacting sentences from your case study prior to publication, not publishing the entire case study publicly or requesting in advance that only certain reviewers be allowed to see the case study (e.g. those with Home Office clearance for research pertaining to national security). This can include the redaction of names, positions and organisations, or entire quotes if necessary, to protect the identity of those who provide testimonials.

Finally, it is worth considering whether you need informed consent and ethical approval before you try and obtain testimonials. This is likely to be particularly important if you actually record interviews as the initial basis of your testimonial. Make sure you and they are clear about whether they are providing a personal testimony or one on behalf of their organisation (ideally the latter as you will want to be able to quote their position in the organisation to underscore the legitimacy of what they have said about your research).

Writing up an impact evaluation as a research article

So, you evaluated your impact and you have the evidence sitting on your desk. How do you make this publicly available, and might there be a win-win situation if you could publish this as part of a peer-reviewed journal article? In an ideal world, evidence to corroborate the impact of your research should be: i) independent; ii) robust and unbiased; and iii) open to scrutiny. However, finding evidence for some impacts is more challenging than for others.

Some impacts are relatively straightforward to evaluate, and are unlikely to take a lot of time. For example, another organisation may be interested in your impact and may collect, analyse and publish data on your impact for you (such as impacts that show up in government statistics, or an organisation that evaluates the impact of your work on their practice and the outcomes they are looking for). In some cases, to credibly claim an impact all you might need is a testimonial from someone at the top of an organisation clearly stating the benefits of your research and that these benefits would not have been possible without your work.

To evaluate and evidence other impacts may require a substantial research effort. For example, in many disciplines, impact evaluation is part of the research process or may in fact be a research project in its own right. Many medical researchers will need a randomised controlled trial to be able to credibly evidence the impact of the treatments that arise from their research, and detailed monitoring and evaluation is typically built into international development research projects. For other impacts, you may want to draw on social science expertise to design an evaluation that unambiguously demonstrates cause and effect and attributes impacts to your research.

To make this happen, you have a number options open to you. You could commission an independent consultant to design and carry out the evaluation for you. However, despite the fact that the evaluation is independent, it is unlikely to appear so on a list of 'sources to corroborate the impact' in an impact case study if it is published on your university website. The alternative therefore is to work with a stakeholder or public representative organisation linked to your impact case study to co-design, carry out and publish the evaluation (potentially with assistance from social scientists in your institution). If they do not have the resources or expertise to carry out the evaluation themselves, then you might want to pay for the independent consultant to do the work for them. They will have to acknowledge your funding, but the evaluation is independent and it is published on their website, so it *looks* independent as well.

Not everyone has access to the funding to make this possible, however. A further option therefore, which requires time and expertise instead of funding, is to conduct the evaluation yourself with social scientists in your institution and write it up as a peer-reviewed paper. In some disciplines, it may be possible to build an impact evaluation into the next paper you write that builds on your underpinning research. This may even add value to the paper, giving it greater academic impact. For some, it may be possible to write the evaluation as a stand-alone paper for a journal in their discipline. For other disciplines, this is not an option, and the researcher who has generated impact is reliant on working with social scientists to co-design, carry out and publish the evaluation in journals from their own field. The challenge here is to pitch this as an opportunity, as there are few social scientists who are likely to want to spend significant amounts of time providing evaluation services to academics from other disciplines. However, with some exploration and discussion, it may be possible to get a social scientist excited about the research opportunities presented by an evaluation.

Although not independent, if you can design your evaluation robustly and get it published in a reputable peer-reviewed journal, it is unlikely that your evidence will be dismissed. In theory you might have 'cooked' the results to make yourself look good, but a strong research design in a good journal is likely to reassure most panel members that you're not trying to pull the wool over their eyes. Here are three types of impact evaluation leading to papers that I've helped design.

As Director of Engagement and Impact for the School of Natural and Environmental Sciences at Newcastle University, I helped one of my colleagues design an evaluation study to corroborate his impact. He approached me with evidence that a paper he had written had led to a massive spike in the consumption of a product across Europe. Sadly, the evidence was only in the form of a correlation, and did not prove cause and effect. The spike in consumption was significant and sustained, and coincided perfectly with the publication of his findings, but how could we demonstrate without doubt that his paper was the cause of this spike rather than a coincidence?

He explained to me that he was planning to publish another paper with even more striking and far-reaching findings early the next year, and we identified an opportunity to do things differently. This time he would go back to his funders to commission a large-scale survey of consumers in the UK and Germany immediately before and after the publication of his work. We would ask them about their consumption of the products he was studying, and whether any changes we detected after publication were in fact as a result of reading about his work. I pitched the opportunity to a social science colleague of mine, Lynn Frewer, who jumped at the opportunity to work with us to test some hypotheses she was developing about consumer behaviour. Together we designed a methodology that should be sufficiently robust and novel to get our findings published in a good journal.

In a second example, I evaluated the impact of my own research with interdisciplinary environmental scientist Ros Bryce (University of the Highlands and Islands), funded through an EU project concerned with science-policy dialogue. We conducted interviews with members of the policy community, asking them to identify whether they had come into contact with over 70 research findings (one of these findings was a paper I had written with colleagues). We then used qualitative interviews and Social Network Analysis to trace those findings as they were communicated from person to person until they either reached policy or not. Life intervened and we didn't finish writing the paper, but in the intervening period, I met a human geographer, Ruth Machen (Newcastle University), who was studying similar questions with the same group of policy-makers. The resulting paper in *Evidence & Policy* combines insights from the qualitative and quantitative analysis that Ros and I conducted with Ruth's in-depth work. It makes a significant methodological contribution in addition to the new empirical understanding of the specific science-policy community we studied.

Policy report to research article

The second type of impact evaluation I've written up as a research article started life as a policy report. This example is more of an attempt to evidence rather than evaluate policy impact. I was faced with a problem whereby I had invested significant research effort to support a policy document (the United Nations Convention to Combat Desertification's Global Land Outlook), but I did not have any control over the way our research would be cited in their report. I needed to create a publicly available evidence trail that would prove that a significant part of the report arose from my research.

To do this, we wrote the first draft of our contribution to the report as a research paper, and then converted this into simpler language for inclusion in the report with a request that it cite the research paper that we then submitted to the journal *Land Degradation and Development*. The journal article includes a footnote on the front page explaining, "a condensed version of this paper is part of the UNCCD's Global Land Outlook, published 17 June 2018". The idea was that whether the report cited our work or not, we would have a published link between the two outputs, with clear overlaps in the text and ideas contained within the two documents. In the end, the policy document cited a working paper version of our paper with the same title, and the link is robust.

Bolt on

Finally, I published a paper in *Global Environmental Change* that contained a substantive section detailing impact evaluation work I conducted with PhD student Kathleen Allen. We did the work in collaboration with the IUCN UK Peatland Programme, which funded her travel expenses to accompany me to run focus groups across the UK to evaluate the impact of research I had submitted to REF2014. The impacts were mixed, but made for interesting reading, enriching a paper that would otherwise have been more descriptive in nature. This enabled us to publish the work in a higher-impact journal than might otherwise have been possible.

References:

- Thomas RJ, Reed MS *et al.* (2018) Modalities for Scaling up Sustainable Land Management and Restoration of Degraded Land. *Land Degradation & Development.*
- Reed MS, Bryce R, Machen R (2018) Pathways to policy impact: a new approach for planning and evidencing research impact. *Evidence & Policy.*
- Reed MS *et al.* (2017) A Place-Based Approach to Payments for Ecosystem Services. *Global Environmental Change* 43: 92-106.

Evidencing international policy impacts

Linking different types of policy impacts to specific pieces of evidence from research can be a major problem for researchers who want to evidence the impact of their work. This attribution challenge is amplified significantly if you are working with overseas governments. It is greater still if you think multiple governments around the world may have taken up your work. How do you find out if your work is being used in government policy, let alone evidence it?

Some types of policy impact may be harder to evidence than others, and you may need to take a different approach to evidencing each type:

- Research publication cited in policy-informing document that (might) be used in informing a new or changed policy
- Actual new or changed policy
- Effect of that new or changed policy being put into practice, with measurable benefit/reduction of harm that has resulted

To find references to research in policy documents and collect evidence of actual policy change, the first step is an Internet search:

1. First, narrow your search down to countries or international organisations (for more global issues) you think are more likely to have used or benefited from evidence from your research (for example, based on the prevalence of the issues you address in your work).
2. If this is a long list of countries, rank according to those you think are likely to be able to benefit most (e.g. prevalence of a disease your work is targeting or share of international greenhouse gas emissions from a sector you are targeting).
3. If there is a phrase or statistic linked to your work (e.g. the name of your method or intervention, or a specific number based on your analysis), use these as search terms.

An alternative way of doing this is to use Altmetrics to identify documents linking to specific research papers. For a guide on how to do this, see Box 14 and Figure 22.

Box 14: Using Altmetrics to find evidence of policy impacts

In some cases, Altmetrics can find evidence that your work has been cited in policy documents for you. This is how to find out:

1. Navigate to your research outputs online.
2. Download the free altmetrics add on ('Boomarklet'): https://www.altmetric.com/products/free-tools/bookmarklet/.
3. Highlight the DOI.
4. Click on the Altmetrics bookmarklet in the toolbar and a small window will open showing the Altmetric data.
5. Click on this to obtain full results.
6. Citations in policy documents are indicated in purple in the 'Altmetric wheel'.
7. Clicking on the policy tab gives the list and sources of the policy documents identified by Altmetrics.

Note that if you search through Scopus and click on a result, there is a link on the right-hand side of the page to associated Metrics. However, at the time of printing, although other Altmetrics data is picked up, the policy metrics are NOT indicated.

Examples of policy sources covered by Altmetrics:
- UK's gov.uk policy papers
- The Publications Office of the European Union
- Intergovernmental Panel on Climate Change (IPCC)
- National Institute for Care and Health Excellence (NICE) evidence search (UK)
- Médecins Sans Frontières (Doctors Without Borders)
- AWMF — Association of Scientific Medical Societies (Germany)
- International Committee of the Red Cross

- European Food Safety Authority (EFSA)
- Intergovernmental Panel on Climate Change (IPCC)
- International Committee of the Red Cross (ICRC)
- World Health Organization (WHO)
- International Monetary Fund (IMF)
- Oxfam Policy & Practice
- UNESCO
- World Bank
- Food and Agriculture Organization (FAO) of the United Nations
- National Academies Press, USA
- Australian Policy Online

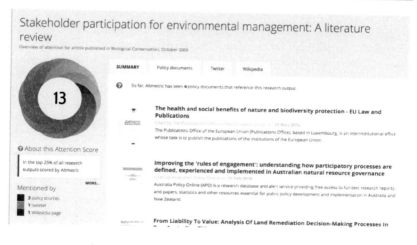

Figure 22: Screenshot of Altmetric results for Reed (2008), showing policy documents citing the paper

If a targeted search for evidence of impact in these countries does not work via a search engine, then this does not necessarily mean there was no impact. The work may have informed policy but without citing your work or any identifiable number or phrase that might help link the policy to your work obviously. At this point you have two options:

1. Systematically look through recent legislation published on the government's website to find policies that address issues linked to your research that were passed into law after your findings were published, and search for any evidence of potential influence. These clues will need to be corroborated through interviews; and

2. Conduct telephone interviews with members of the civil service and wider policy community to identify policy impacts. Getting names of relevant people can be tricky, given rapid staff turnover in most organisations, and response rates to emails requesting information are often poor, so the most efficient approach may be to pick up the telephone. Go from person to person till you find the most relevant teams and people. In some countries, it can be easier to identify and speak to others from the wider policy community of people who regularly interact with policy-makers, for example, charities and think-tanks.

If this still fails to uncover any evidence of international policy impact, then you can avoid wasting your time by using the contacts you make to start promoting your evidence. If you start your investigation with the goal of helping further inform policy, then you may get better responses to your initial enquiries, compared to just asking for evidence that your work has been used. Consider whether there are ways you can add value easily, for example, by turning your work into a policy brief, manual or handbook that can make it easy for members of the policy community to act on your evidence and learn from others who have implemented it already elsewhere. If this is difficult to do directly, then research the broader context in which policies are being developed in the country to identify organisations that may be trying to influence policy in areas relevant to your work, and consider whether you could empower them to achieve impact via your research. These organisations can be powerful allies who can affect change on your behalf, but tread carefully. Some organisations you work with may cherry-pick or distort findings to meet their goals. Depending on the context you are working in, aligning yourself with organisations that are trying to influence government policy can risk your personal safety as well as your academic and personal reputation. Indeed, some researchers go so far as to deliberately not engage with any external groups who might have an interest in their research, so that they can maintain their independence and academic integrity.

Even if you have identified impacts already, it is always worth asking whether you can add to these impacts, for example, by working with agencies that are charged with enforcing policy to ensure that the policy is implemented effectively on the ground in a manner that is consistent with the evidence you have generated. Your research may have been discussed in parliament, a committee or have been cited in a policy document, but rather than leaving it there, you may be able to advise the government on how they might be able to actually turn this into legislation that can effect real change (especially if you have already learned how other governments have done this successfully). It is typically easiest to find junior civil servants working in relevant evidence or policy teams, at the bottom of the organisation's hierarchy. If it is not possible to find details of relevant people online, it may be possible to go via telephone switchboards, or via in-country academic partners who have relevant contacts.

Collecting evidence of impacts arising from policy implementation is much harder. In many countries, it is normal practice for civil servants to conduct policy reviews. These aim to assess whether the policy met its goals or targets were reached. University researchers typically have the credibility and independence necessary to help with these reviews, and if an offer of help is accepted, you may be able to play a key role in obtaining the evidence you need.

How to set up a stakeholder advisory panel for your research project

Projects that engage with stakeholders via advisory panels are perceived to impact positively on stakeholder knowledge, policies and practices as well as improve research quality and relevance (see Chapter 6 for more details). I have used stakeholder advisory panels in most of the projects I have led, and have found them invaluable. With help from members of advisory panels, I have found new and better pathways to impact which would otherwise not have been available to me. They have helped with the research, providing me with new insights and interpretations, access to data and new funding opportunities. In some of my social science research projects, I have also been able to use the panel as a focus group to triangulate and increase the robustness of my findings. In other cases, I have been able to identify marginalised groups with similar interests to more powerful groups, and use the advisory panel as an opportunity to foster strategic alliances that empower the marginalised.

This is how you can set up a stakeholder advisory panel for your research project:

- **Scope out who should be on your panel:** do a publics/stakeholder analysis to work out who you want to invite to join your panel (I've explained how to do this in Chapter 14). There is no single correct way to use the findings of your analysis. You may want to prioritise those who will benefit most or who have greatest influence to help you achieve impact. However, you may also want to ensure you represent hard-to-reach groups and give voice to marginalised groups who may have alternative and valuable perspectives to offer.
- **Choose a chairperson:** many researchers simply give the chair to the Principal Investigator on the project, however, this can be problematic as the researcher on the project is not independent and is less likely to be held to account by the group. Finding a well-respected and competent, independent chair can empower stakeholders to ask probing questions and genuinely shape the research. Giving the chair role to someone

else also frees up the researcher leading the project to genuinely listen and respond to feedback (and many experienced chairs are better at the job of chairing than the project leader would have been). The chair needs to be well respected by everyone on the panel and so will typically be chosen from outside academia, but you may be able to find a highly engaged researcher who is well known in the stakeholder community who can perform the role effectively.

- **Further scope out panel remit and membership:** conduct a small number of informal scoping interviews with your chair (giving them some joint ownership of the process) and key stakeholders to check that there's no one important missing from your invitation list. During these conversations, you can also discuss the role that the group is going to perform for your project, and find out whether you need to offer a non-disclosure agreement (NDA) to enable everyone to speak freely during meetings (typically for panels involving stakeholders from industry or on highly controversial topics). If you need an NDA, speak to the legal team in your institution to get something drawn up.
- **Keep numbers manageable:** for a stakeholder advisory panel to operate efficiently and enable genuine discussion, you should aim for somewhere around 8-12 members. If you have over 20 members, meetings will become difficult to manage and there may not be time for everyone to express their views and share their knowledge. If there are compelling (e.g. political) reasons why you need to engage with a large group intensively and regularly throughout your project, consider supplementing your advisory panel with a wider 'reference group' (of any size). Reference group members get regular updates via email (in addition to any wider newsletter you might publish), giving them privileged access to your work and the opportunity to input via email and one-to-one calls as necessary. This avoids key groups feeling excluded whilst retaining a manageable size of advisory panel.
- **Invite members:** turn your research proposal into a project flyer using plain language and emphasising the impacts you hope to achieve, and send this along with your invitations. Tailor your invitation to the interests of different groups based on your publics/stakeholder analysis, emphasising the aspects of the research you think are likely to be of greatest interest to them. An additional step after this for some projects is to issue non-

disclosure agreements in advance of the initial meeting so everyone has time to scrutinise and sign the agreement ready for the first meeting.

- **Hold your first meeting:** make sure you cover the following points in your first meeting:
  - Make time for introductions, giving everyone as much time as the research team to introduce their interests and expertise to the rest of the group.
  - Explain your research in plain language, providing ample time for questions and discussion.
  - Agree your Terms of Reference for the group: ideally come prepared with a draft and send this in advance as one of the papers for the meeting. I've copied an example Terms of Reference for an advisory panel below.
  - Manage expectations: although you are providing panel members with an opportunity to advise and shape your research, there will be limits to how much you can adapt your plans, and people need to know this.
  - Provide networking opportunities: if you did your publics/stakeholder analysis well, you should have identified people from different backgrounds who share similar interests, and one of the things they will benefit most from initially is the opportunity to network and learn from others in the meeting. Therefore, make sure you provide plenty of time for breaks, and consider whether you can offer a lunch before or after the event to provide people with more protracted opportunities for engagement.

Example Stakeholder Advisory Panel Terms of Reference

Purpose of the group

The Stakeholder Advisory Group (the Group hereafter) will:

1. Ensure project goals are consistent with the needs of beneficiaries, suggesting where feasible, additional work to help realise social, environmental and economic benefits for the broadest possible range of stakeholders and publics
2. Review and provide feedback on project progress towards stated goals

3. Feedback on scenarios developed in the research and, when available, take part in 'social innovation labs' to generate new ideas for testing in the research
4. Contribute towards development of options for upscaling project findings across the UK and internationally in the XXX sector and to other sectors

Governance and roles

The Group will be chaired by Prof X (Dean of XXX, University of XXX). Prof Y (Principal Investigator, XXX University) is responsible for reporting project progress to the Group and progress against recommendations from the Stakeholder Advisory Group at meetings and as necessary between meetings. Work Package leaders are expected to report to the Group, if possible in person, to provide progress reports on their work and answer questions from the Group. Prof Y is responsible for reporting on the activity of the Group to the XXX programme secretariat at the University of XXX. Group members may deputise within their own organisation. Annex 1 details membership of the group. Membership is by invitation only. The Chair will adjudicate over any proposals to add or remove an organisation from the group, and his/her decision will be actioned by the project.

Meetings

The Group will meet annually, but can request additional meetings of the whole group or sub-groups as necessary. All group members can request an individual meeting with Prof Y to discuss issues pertaining to the project at any time by arrangement. Members may join in person or via video link. Meeting minutes will be taken by Project Manager, XXX.

Confidentiality

All group members will be asked to sign a non-disclosure agreement prior to the start of the first meeting. See draft agreement for details. The names, positions and organisations of Group members will be publicly available on the project website. Minutes will only be circulated between Group members.

Communication between group members between meetings will be primarily via email from the project. Contact details of Group members will not be disclosed unless permission is given. As core project partners, project materials may contain branding from Universities of X, Y and Z, A, B and C (in addition to funder logos). Logos of Group member organisations will not be used on project materials.

How to turn your next paper into an infographic

Infographics are a great way of communicating your research to a wider audience, from other researchers to people who might use your research outside academia. A range of websites now exist to help you develop your graphics, but many researchers get stuck trying to come up with ideas to visualise their research findings, and then, once they've produced their infographic, fail to get it to the right audience. In this guide, I will explain how you can turn your next paper into an infographic in six simple steps. Below, you can see examples of how I've used these six steps to develop infographics for researchers. Although I like to work with a designer, you don't need one to turn your research into an infographic.

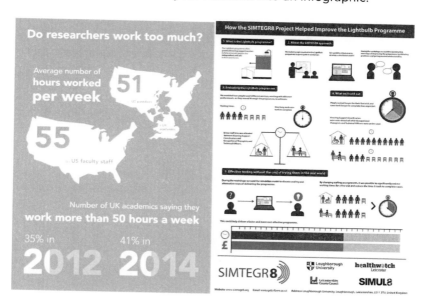

1. Extract your key messages

Start by cutting your article down to size. If the abstract doesn't summarise your key points, extract them from the work as short phrases, similar to the bullet lists of 'highlights' many journals now ask for. Depending on your research, you may also need to have a phrase you can visualise to summarise the problem you are seeking to solve or the aim of the work, and the methods you used.

2. Simplify your language

Cut the academic jargon and rework your messages to use as few words as possible, making sure that they are instantly understandable. Even if your audience is other academics in your field, people need to be able to scan the limited text in an infographic rapidly to get your message, and the simpler your language, the more effective your infographic is likely to be at communicating your key messages.

Upgoer6 (http://splasho.com/upgoer6/) is a simple text editor that will colour-code your text, showing you how commonly the words are used in the English language. The word turns green if you are using one of the 1000 most commonly used words. If it is red, you might want to think about finding simpler language.

3. Visualise your key messages

For some people this comes easily, but for most researchers this is a real challenge, so here are a few ideas that will help get the creative juices flowing:

- Draw all the images that come into your mind as you think about each point, choosing the images that you think communicate the point most powerfully.
- Do a Google image search for inspiration.
- Turn to a thesaurus to search for synonyms of key words that might have the same meaning but are more instantly visual.
- Put key words into the NounProject (https://thenounproject. com/) to get ideas for icons that express your ideas visually in simple terms.
- Talk to someone about the ideas you are trying to communicate to clarify your thinking and explore options for visualisation.
- Get a second opinion — what seems an obvious representation to you may not be obvious to others.

4. Come up with a layout

Depending on how you want to communicate your infographic, you may need two alternative layouts: a long and narrow layout with your title in the middle for Twitter (so the title shows in the preview

338

pane of your tweet); and a landscape version that you can use in presentations and publications. Draw your layout on a whiteboard so you can play around with different versions till there is a logical flow of ideas for the reader to navigate through.

5. Convert to graphics

Find and pay for copyright to use stock images and use infographic template websites like Easel.ly, Piktochart, Venngage, Visme, or get a designer to help you with this stage. A simple alternative is to create a series of infographics that have a stand-alone message that builds over the series. It can be as simple as finding a striking image and overlaying text using an image manipulation website like PicMonkey. If you create a set of four images, you can release them one per day, and then include all four in a single tweet on day 4.

Don't go crazy with the colours. If you're overlaying text onto a photo, pick a colour from the photo that contrasts with the majority of the image. If you're creating your graphics from scratch, use a limited palette of between 2 and 3 colours (not including your font colour) that complement each other well. Pick a simple font so that it's easy to read, and try to use colours that will match the tone/content of the text, e.g. dark blue is formal and calm, and green is commonly associated with environmental topics.

6. Have a plan for communicating your infographic

There is no point in making an infographic if you don't have a plan for getting it to your target audience:

- Know who your audience is, what they are interested in and what platforms they are on.
- Decide what you want people to do as a result of reading your infographic, and make it easy for them to do this. If you want other researchers to cite your paper, link directly to the open-access version of the full paper. If you want a wider audience to engage more deeply with your work, consider linking your infographic to a blog, which then links to the original paper. If you want someone to perform an action, facilitate this, for example, by giving them links to alternative products or a draft email to contact their local MP.
- Create a social media strategy (it's not as hard as you might think — you just have to be able to answer four questions, in

your head). For help, see my Social Media Strategy template (later in this section of the book).

- Alternatively, save time and reach out to online influencers who already have a large audience that you want to reach. Reach out to them on the platform, and if they don't respond, email or phone them, explaining why your infographic will be of interest to their audience and add value to what they are doing. You'll be surprised how many will agree to work with you if you are persistent.

How to write a top-scoring impact case study for the UK Research Excellence Framework

Three key lessons

1. Articulate and evidence significant impact

   - Adjust interpretation and expectation by Unit of Assessment (UoA):
     - Applied research: more likely to focus on benefits for economics, environment, policy, health/well-being, and other forms of decision-making and behaviour change via stakeholder engagement
     - Pure sciences, arts and humanities: more likely to focus on benefits in terms of awareness, attitudes and cultural change, and feature wider public engagement than applied research
     - Look at the publicly available data from REF2014 to see examples of impact case studies in your UoA that scored highly for inspiration, but be aware that this can only be a guide, as panel members will differ and panels may set new norms in REF2021.
   - Create a problem statement to frame the significance of the impact.
   - Cut out anything that looks like a benefit to researchers, the academy or your discipline.
   - Look for additional types of impact you may not have originally planned for in case these are significant, e.g. an innovation that saves money by reducing the energy intensity of production may have as impressive an environmental impact as the economic impact that was originally sought. Consider if you might have missed the following types of impact:
     - Awareness and understanding
     - Attitude change
     - Economics
     - Environment
     - Health and well-being
     - Policy
     - Other forms of decision-making and behaviour change
     - Cultural change

- If you have capacity-building or awareness and understanding impacts, ask whether it is possible to turn these impacts into any of the other impacts above within the time available to you.
- Benchmark your numbers: 1000 Twitter followers or £100K of revenue might seem impressive to you but might look embarrassing next to competitors. Have a look at the sorts of numbers claimed by 3* and 4* case studies from your UoA in 2014 to get a feel for the sorts of numbers that might actually look impressive.
- Don't overstate your significance: if the experience was 'transformative' for participants, make sure you are able to articulate and evidence exactly what was transformed.
- Consider integrating case studies working in similar areas that are graded as 2* in your internal review where you think they might get 3* if combined.
- Unless you are pursuing a case study based on public engagement (where it could be a useful indication of reach), media work is usually only a pathway to impact, not evidence of significant impact in itself.

2. Provide evidence that impacts are far-reaching

- Adjust interpretation and expectation by UoA:
  - More examples of global reach in some areas, e.g. clinical medicine, public health or earth systems and environmental science
  - More examples of high-scoring case studies with sub-national reach in other areas, e.g. sociology, anthropology and English language and literature
  - As above, high-scoring case studies from REF2014 can give you a sense of the norms for reach in your UoA.
- Frame the case study with a problem statement that is aligned with the scale at which impacts can be evidenced:
  - Don't use an international framing for a national or sub-national impact.
  - Explain why national or sub-national impacts address an important need that is uniquely felt at that scale.
  - If the majority of the evidence is national or sub-national and you also have some more limited evidence of international reach, keep the national or sub-national framing and emphasis, and refer to wider interest more briefly to emphasise reach without overstating it or

undermining the core impact by using this wider interest to frame the case study.

- Look for additional beneficiaries you may not know are using or benefiting from your research. Reach can be on different geographical or social scales, or around communities of interest that may cut across geographical and social units.
- Systematically identify groups, sectors or countries that have similar problems or needs in similar contexts that might benefit from your impact.

3. Submit the impact, not the pathway to impact

- Ensure that the majority of the text in the 'summary' and 'description of the impact' is about the impacts rather than the pathways to impact.
- To ensure you are submitting the actual impact, and not just the pathway to impact, keep asking "what was the benefit and why was this important?" and describe the benefits more than the process through which those benefits were derived. If you don't know why it was important, ask the beneficiaries to tell you what was meaningful or valuable to them.
- Dissemination is not impact: even if you have impressive numbers of reads, downloads, views or listens, how do you know if anyone learned anything from it, benefited, or did anything different as a result? Keep asking "what happened next?" until you find the benefit. To do this, design your communications so you can legitimately follow up longitudinally with audiences to re-engage, deepen interest and learning, and ask them how they benefited.
- Developing resources for schools and doing work in schools is a pathway to impact, not an impact in its own right. Identify specific changes you would like to see (e.g. increased attainment in a specific subject, reducing an attainment gap between boys and girls or ethnic groups for a particular subject, or influencing choice of subject at university) and use your materials and work in schools as interventions designed to achieve these impacts. Follow up to find out whether your interventions worked.

343

- Make sure the majority of the words in your summary are about the impact, rather than the context and pathway to impact:
  - Spend time making sure your summary resonates strongly with readers, communicating your impact straightforwardly and persuasively.
  - Get multiple reviews of your summary and keep polishing till you can improve it no further.
- Target your case study to the appropriate UoA based on the publications in the 'underpinning research' section:
  - Case studies submitted to UoAs by researchers from outside the discipline because they think assessment criteria will be easier may find a less sympathetic or understanding audience than they expected.
- Make sure the underpinning research reaches the 2* quality threshold, ideally based on the judgment of two independent assessors:
  - Being published in a peer-reviewed journal or having a book deal with a prestigious academic publisher doesn't guarantee 2* status if it is weak work.
  - Outputs that are not peer-reviewed or academic books, such as final reports to funders, can reach the 2* threshold if they are strong work and publicly available.
- Describe the key findings from the underpinning research that pertain to your impact:
  - Only include essential contextual material, and avoid unnecessary detail on methods or other findings that were not integral to the impact.
  - Number your list of underpinning research outputs and cite each output by number in your description of the underpinning research.
- Ensure your corroborating evidence is: i) independent; ii) robust; and iii) publicly available:
  - Work with credible stakeholders to conduct independent evaluations of your impact, e.g. by stimulating a policy review or making the case to a project partner that their stakeholders might be interested to know the impact of the work you did with them.
  - Where resources prevent this, consider providing your stakeholders with funding for an independent consultant to

344

evaluate the impact on your behalf, acknowledging the source of funding in the report that is published on the organisation's website.

- ○ Offer help designing the evaluation to ensure it is robust.
- ○ If you can't find an organisation to do this for or with you, do it yourself and publish the findings in a well-respected peer-reviewed journal. The evaluation will not be not independent, but if well designed and written, few reviewers will doubt the veracity of your claims.
- ○ Although some impacts will require rigorous research to prove cause and effect beyond reasonable doubt (e.g. randomised controlled trials to demonstrate the efficacy of an intervention, or social science studies of attitude or cultural change), the evidence just needs to be credible and convincing enough, which in many cases will be less resource intensive. Take advice from your UoA lead if you are worried that the effort required to evaluate your impact is disproportionate, as they will have a more strategic overview of the number of case studies needed and their maturity, and can help you decide if the effort is worth it in the context.
- Assess impacts longitudinally to look for longer-term benefits that may not have been apparent at the time you did the work:
  - ○ For policy impacts, where possible include evidence that the policy was implemented, enforced and worked on the ground.
  - ○ For public engagement impacts, create opportunities to re-engage with participants to deepen their interest and learning and find out how they have benefited in the long term.
- Make sure case studies that were previously submitted to REF2014 only claim additional, new impacts that took place after 2014:
  - ○ You will need to declare if your case study is a continuation of a REF2014 impact. Rather than leaving panel members to decide whether or not your case study is additional, make this case explicitly.

- Give your most significant, far-reaching and well-evidenced impacts room to shine:
  - Don't crowd out impressive impacts with long descriptions of the pathway to impact, or descriptions of linked impacts that are less impressive.
- Choose your words carefully:
  - Low-scoring case studies bear more resemblance in written style to the underpinning research they were based on, including technical language and jargon, and tend to be less specific in their use of language, e.g. describing poorly specified improvements rather than describing actual change.
  - High-scoring case studies are typically written in concise, accessible language, making specific claims with clear causal links back to the underpinning research.
- Tell a story:
  - Create a clear narrative which respects the timeline and realities of the pathway to impact, but which builds logically and sequentially towards the impact as the culmination of the story.
  - Link disparate impacts together into a coherent narrative using thematic links if there were not links in practice.
- Use structure to your advantage:
  - Signpost each of your key impacts so they cannot be missed.
  - Systematically explain why each impact is significant and far-reaching, referring to evidence.
  - Break up the text to make it easier to read, using sub-headings and other structural devices.
- Use context to add shine to your pathway to impact and evidence of impact:
  - Put evidence in context to explain its significance, e.g. rather than just citing visitor numbers, say if this was the largest number of visitors in any week that year, or explain who visited and why they are significant (perhaps they are from groups who don't typically visit museums). A small number may be significant in context, but you will need to create the argument to justify this interpretation of your evidence.

- If you worked with a range of organisations, name those that are widely recognised and respected to lend further credibility to your pathway to impact.
- Integrate testimonials to more richly describe and explain the significance of your impact:
  - Ask for quotable, concise summaries of key points (or ask for a draft testimonial, extract the key points as a summary and ask them to amend/approve the summary to be used as a quote). Integrate these quotes with the text of your case study, in case assessors do not have time read the full testimony.
  - Ask direct questions to obtain specific answers about impacts you are interested in, backed up by numbers and evidence where possible.
  - Consider obtaining testimonials via interviews so you can probe to get the most relevant information, transcribe relevant extracts from the interview, requesting any clarifications needed, and ask for a final version on letter-headed paper. Consider whether you need ethics approval and informed consent for this process.
  - Get testimonials from high-profile individuals at senior levels in organisations, but consider also giving voice to low-status beneficiaries, such as school children or patients.
  - Don't wait — request testimonials now. People move on, forget or worse.

Impacts in top-scoring case studies are significant, far-reaching, clearly articulated and convincingly evidenced. While there are examples of 2* and 3* impact case studies that might have scored 3* or 4* if they had been better written and evidenced, it is not possible to get top grades unless you actually have significant and far-reaching impact. There is time to deepen and extend the impact of your research.

Identifying impact indicators

There are a number of ways of coming up with impact indicators. Some will be self-evident, such as the value of a spin-out company or the number of lives saved by a new safety technology. However, even apparently self-evident indicators may be difficult to collect data on, for example, if the company is not willing to disclose profits or your relationship is with the company who make the technology rather than those who use it. These examples show how the process of identifying indicators can actually feed back into your impact planning, as you realise that you will need to initiate new relationships or strengthen trust with key individuals.

Where indicators are not obvious, it may be necessary to use your imagination to come up with examples of the sorts of things you might expect to change or happen if you achieved your impact. For example, if you wanted to change perceptions of Asian cinema in Europe, you might expect to see changes in the categories in film awards and the composition of judges on film award panels. Eventually this might translate into an increase in the number of Asian films shown on television and in cinemas, with increased viewing numbers. However, if viewing figures were your only indicator of impact, you might miss important changes that demonstrate the influence of your research. In this particular case, the person researching Asian film hadn't been looking at awards, and didn't have a plan of activities to get her research to the awards community. Again, this is a good example of indicator identification pointing to missing activities that can measurably increase impact.

So what makes a good impact indicator? During my PhD, I reviewed the literature on indicators, and found that there are a number of characteristics you might want to look for in a good indicator. Depending on the timescales you are working with, and your budget, some of these characteristics may be more important to you than others.

A good indicator of research impact should be:

- **Accurate and bias free:** a good indicator should actually tell you about the impact of your research (or the impact of the activity you designed to deliver impact from your research), rather than just describing the pathway to impact. For example, rather than just measuring the number of people who attend your events, give them a questionnaire to ask how your event changed their perceptions and what they will do differently as a result, and try and get their contact details and permission to follow them up later to see if they put what they learned into practice.

- **Relevant, reliable and consistent in a range of different circumstances:** you don't want to choose indicators that will only give you accurate information in one country or culture if you expect your impact to occur across multiple countries and cultures. Similarly, one measure might work for one public audience or stakeholder group, but not apply to another.

- **Timely:** it is useful if your indicators can provide you with information in a timely manner, for example, within a reporting window for your funder. Also, if it takes too long to collect or analyse the data you need, you may find out too late that something isn't working, and be unable to correct your course.

- **Robust and credible:** consider your indicators from the perspective of the people who will be learning about your impacts, and assess whether or not you think that the evidence you are collecting is likely to be perceived as being robust and credible. You may have designed evidence collection in a way that you know is highly credible, but the simple fact that you are collecting that data as the researcher who is claiming the impact may undermine the credibility of your evidence. Consider if there is some way to get the evidence you need collected and published by an independent, credible third party. Check whether the indicators you have chosen have been used by others and whether they are respected or not. For example, the case study in Chapter 22, 'Discover Turner's Yorkshire', used 'Advertising Value Equivalency' to estimate the value of media coverage. This measures column inches devoted to stories about your research and then looks at the price that the publication charges for advertising space, and uses this to convert the column inches into a financial value.

However, this approach makes many assumptions (e.g. that the whole article is actually about your research, and that the coverage is positive) and as a result is discouraged by the Chartered Institute for Public Relations (CIPR). CIPR offer guidance on alternative, more robust, measures, such as qualitative research via interviews and focus groups, and quantitative research via polls, surveys and other studies that can credibly measure the impact of your work.

- **Independently verifiable and replicable:** good indicators have an evidence trail behind them that can be traced by anyone who wants to understand how you arrived at the numbers you have claimed in relation to your impact. As I mentioned in the previous point, having evidence published by an independent third party helps make it credible. However, this is even better if there is a transparent method that has been used to collect the data which could be replicated if someone wanted to check the figures for themselves.

- **Linked to clear targets or baselines:** it is sometimes difficult to interpret the importance of a change in an indicator if you are unable to see the original baseline level from which it has changed, or how far it has changed towards an impact target. You should already have SMART impact objectives that you're planning for and monitoring progress towards, so make sure that you've found as many indicators as possible that could enable you to see if you are moving towards or away from these objectives. In addition to this, it can be useful to consider your baseline. Very often, to determine the baseline from which your research is building impact, you will need to assess this at the start of your project. For example, if you expect a habitat to be saved or restored as a result of your research, then you will need to know how much of the habitat there is when you started your work, and what condition that remaining habitat is in. This illustrates the importance of identifying indicators while you are planning for impact at the outset.

If you're struggling to come up with measurable indicators, here are some examples of impact evidence you can explore:

- Google key statistics or phrases from your work to see where they have been used in documents that have been put on the web.

- Did people pay for advice or did consultancy work arise from the research, and can you track the flows of money?
- What is the readership of a particular newspaper or online media source and how might this translate into money via advertising revenues linked to page views and circulation? Retain date-stamped copies of the web pages in case the web links change.
- Did particular target groups attend project events? This could involve retaining delegate lists from conferences or workshops (particularly if policy-makers and media attended) and following these up to assess if there was any benefit or change as a result.
- Did you get letters of support from external bodies before you submitted the grant that funded your research? If so, can you go back to these organisations to see how they used your work? Who did they pass your work on to or talk to about your research, and at what decision-making level? Can these people point you to evidence in documents etc. that your work has been used? If not, can they provide testimonials about the impact of your work?
- Did your engagement with stakeholders continue after the project? If so, what further collaborations arose from this engagement that wouldn't have happened without the original research? Did collaborators use the methods from the research in their own professional practice after the research ended, and what was the effect of those changes in their practice?

How to crowdfund your research to engage with the public

As research funding from traditional sources becomes increasingly competitive, many researchers are now turning to the public to directly fund their work. However, with these new opportunities come a number of pitfalls, so you will need to think carefully about whether or not your research is suited to this approach. For me, one of the most exciting things about the crowdfunding model is the way that it drives engagement with your research, helping people understand what you're doing to a depth that is rarely possible through other forms of online engagement.

Crowdfunding websites like Kickstarter and IndieGogo are increasingly funding projects in technology, the arts, campaigns and non-profit community work, with some projects raising significant sums of money. Projects listed on these crowdfunding websites offer investors a range of rewards, based on the amount they are willing to invest, with small rewards available for as little as a £1 investment, rising to personalised products and events to reward larger investments.

Now a number of websites have started to offer opportunities to invest in research, for example, experiment.com and RocketHub. These initiatives are less about getting your hands on the latest gadget before it hits the shops, and more based on the warm glow of knowing you've advanced knowledge and made the world a better place. Having said this, many projects do offer personal rewards.

Crowdfunding research involves raising money directly from the public, with the research project idea articulated on a dedicated project page along with an invitation for individuals to help fund it. Typically, the project only goes ahead once the funding target has been reached, and only at that point are supporters' payments actually taken. Depending on the platform you are using and the way you set up your project, you may still be able to take funds that are pledged if you do not reach your target, but you will need to be able to deliver project outcomes, including rewards, with whatever money you raise.

What sort of projects are likely to get crowdfunded?

The answer to this question depends upon the sort of project you want to fund. If it will lead to a technology, product, experience, performance or some other tangible output that members of the public may want to own or experience, then the chances are that you may be onto a winner. A successful crowdfunded project needs to be beautifully and effectively presented and explained, making it clear how it will make a difference. You'll need a strategy to drive people to your project page and get them interested enough to read about what you plan to research (blogging, tweeting, getting traditional media coverage etc.). To attract visitors, many crowdfunded projects offer affordable and desirable rewards to investors.

Most crowdfunded research projects are relatively small, requiring a few hundred or thousand pounds. Therefore, current crowdfunding of research is often used for 'seed-corn' funding bigger projects, testing ideas and prototypes, which can then attract larger amounts of funding from more traditional research funding sources. However, with a bit of clever project design, researchers can 'think big' when using crowdfunding, and divide projects into self-contained work packages that can function effectively on their own. Then, if more than one work package is funded, they will link together to form a larger, longer-term project.

Importantly for us as researchers, crowdfunding our work can also open the doors to better public engagement and communication. Crowdfunding platforms represent an emerging form of collaboration between researchers and those who use our research, where the public can be informed and inform the research process. For example, FundaGeek enables discussion forums where the public can debate the value of a project alongside any moral concerns.

Tips for a successful crowdfunding campaign

If you think you've got an idea that might fit the crowdfunding model, then here are a few tips that will help you make your research idea become a reality:

1. Take a look at successfully funded projects to get inspiration for the sorts of rewards you might be able to offer investors, from

highly affordable options to more expensive options for larger investors. We recommend that you look at projects that have been successful in areas with a longer history of crowdfunding, e.g. technology, charity and the arts, to get ideas about what works.

2. Purge your project of all jargon, so you can communicate your project in a way that excites and inspires people to fund you — something most researchers need to practise.

3. Creating a slick video to promote the project is a must, to make it as accessible as possible to potential investors (many crowdfunding websites require this anyway). Make sure your presentation and message are personal, visual, original/unexpected and, if possible, appeal to people on an emotional level (Box 12).

4. Set a realistic funding goal, dividing large projects into self-contained, smaller projects (effectively work packages) that are more likely to get funded.

5. Pay particular attention to creating a social media strategy with clear aims to promote your campaign (Chapter 17). Systematically consider who your audience is, their preferences and interests, the sorts of media and communications they are most likely to respond to, and the offer/reward that people in your audience are likely to respond to. Build your social media following before you launch your campaign, and think about how to target people who are well connected to your target audience (e.g. with large Twitter followings), so they can promote your campaign for you. Communicate your project as widely as possible via social media, and then via your personal and professional networks.

What problems might I encounter?

There are a number of practical and ethical challenges to crowdfunding research. First, it's possible that not all investors will honour their pledges, and so the eventual sum raised may fall short of the target. Although websites take payment details electronically from investors, credit checks are often performed many months prior to the end of the campaign when payments are actually taken. This may mean it is not possible to fulfil your commitments to those who do honour their pledges, and in the worse-case scenario they may demand refunds or sue the project.

Second, not all crowdfunding platforms have a peer review or ethics review process to screen projects before they go online. We can all think of projects that we thought were watertight when we submitted them that came back from peer review or ethics review with fundamental problems that meant they were unfundable. In this case, you may discover these flaws only once you've started spending on the project, and you may not be able to bring it to completion. Even worse, such projects may inadvertently put members of the public at risk and lead to unintended consequences. Take a look at the policies of different crowdfunding websites, and use sites that have some sort of review process. This will also mean your project isn't sitting beside pseudo-science projects that you would not want to be associated with. Particularly useful are sites that include discussion forums where you can engage with potential investors to debate the validity and ethics of what you are doing. In this way, engaging with crowdfunding actually has the potential to enhance the quality of the research you develop. It is also always wise to get feedback from friendly colleagues before you submit anything for funding, whether to traditional funders or a crowdfunding website.

Third, some people worry about the fact that there is no way of vetting the people who fund their research, or knowing if they may have a conflict of interest — for subjects where it is necessary to declare funders and any conflicts of interest as part of the publication process. Although this is something to think about, I have not heard of any projects that have fallen foul of this at publication stage, and there is a contrary argument that the number of different people funding your research immediately demonstrates the perceived value and likely impact of your work.

Finally, before going down this route, make sure you check with the funding office in your institution whether they are able to accept crowdfunding for your research, or allow you to accept this funding for research conducted under their auspices.

So should I crowdfund my next research project?

Obviously it depends on the type of project you are trying to fund, and how easily you think it will resonate with the public. The answer will depend to a large extent on the sort of research field you are working in, and the likely risks of the project going wrong, either in terms of the research itself or health and safety. It might feel risky to try this for the first time, but why not start small with a low-risk, low-cost project that can lay the ground for something more ambitious later?

This new funding model may be particularly attractive to early career academics and those who are good at communicating their research to the public. Unlike many traditional funders which require you to have a permanent or 'tenured' academic post before they'll even look at your funding application, anyone can apply for crowdfunding. You can potentially be up and running with your new project in a few months, compared to the long and drawn-out traditional review process, giving you the platform you need for the bigger projects that will get you greater job security. Crowdfunding may also be increasingly popular with more established researchers, who are seeking new ways to engage with the likely users of their work and generate impact, given funders' increasing focus on the impact agenda.

Crowdfunding research is in its early stages. It is therefore unsurprising that its many potential benefits have largely been overlooked by the research community. This approach isn't all about getting money for research; it is a new way of doing research in collaboration with your funders — the public — giving you an instant route to public engagement in your work. For me, crowdfunding is about the democratisation of science and making sure your research resonates with the wider interests and needs of society.

How to turn your research findings into a video that people actually want to watch

It seems that everyone is making videos about their research these days, but if you look at the number of views, not many of these films are actually getting their message across to large audiences. So how do you make a video about your research that people will actually want to watch?

I think there are two crucial things you need to get right that most researchers overlook:

- Come up with a powerful idea that can act as a vehicle for your research findings, for example, a real-life story or a striking, humorous or thought-provoking metaphor. Make your idea personal, unexpected, visually striking and visceral (Box 12).
- Have a strategy in place to drive traffic to your video. Just having a great video isn't enough — people have to know that it is there. For online videos, social media is a powerful means of driving traffic, so make sure you have a social media strategy in place to harness its power to get your film noticed.

If you have sufficient budget, hiring a professional film company to make a short video about your key findings can be a powerful and highly professional way to communicate messages to policy audiences, as well as to other key stakeholder groups. If you're working on a much smaller budget and cannot afford to pay for a professional video to be made, you may be surprised at how effective it can be to create your own videos, with just a few tips to help make it come across effectively.

I've got no budget — can I make the film myself?

With the low cost of digital video equipment (and integration of video recorders of sufficient quality for online streaming on most mobile phones nowadays), combined with the ready availability of free and easy-to-use video-editing software (such as Windows Moviemaker™ or Apple's iMovie™), producing your own video content is now within the reach of even the most ardent technophobe. Here are some pointers to make the process easier:

1. Plan thoroughly and write a script — this will ensure you get the shots you want and you don't video more than you need, thus making editing much easier.
   a) Spend time thinking about your story, and tell it like a story with a clear beginning, middle and end.
   b) Think about how you can make that story personal in some way to the people who will watch your film, which will make them apply your research in some way to their own lives.
   c) Try and think of something that will take people by surprise — this is one of the elements of a video that is most likely to make someone share the film with their social network.
   d) Come up with some memorable visuals, e.g. some sort of visual metaphor that sums up your research findings, a spectacular location or something entertaining that will help the key ideas stick in people's heads.
   e) Think about how you might be able to engage with people's emotions on some level (ideally positive rather than negative emotions).
2. Turn your script into a 'story board' — little sketches that convey what will happen visually for each section of your script.
3. Pay attention to the sound — if possible use an external microphone for interviews, or make sure the speaker is near enough to the camera's built-in microphone, and watch out for background noise.
4. Be aware of what the person in front of the camera is wearing — some colours interfere with the white balance and exposure, especially if you let the camera do it automatically. Avoid pure reds, whites and blacks, dangly or flashy jewellery (this doesn't just interfere with colour but also sound) and complex patterns or stripes.
5. Always use a tripod for filming static shots and avoid zooming or moving the camera around unless it is absolutely necessary.
6. Make the editing software work for you — use titles, transitions and effects to convey meaning and make your video look more polished, but beware, over-using effects can be distracting and may look unprofessional.
7. Get clearance — getting signed consent forms from participants and using only copyright-cleared materials for things like images and soundtracks could save you massive potential headaches later on.
8. Make videos available in as many formats as you have time to create in order to improve accessibility (e.g. YouTube, Vimeo,

podcast, embedded in your project website, links to download files in .mp4 and .wmv formats).

9. The optimum length of a video on YouTube is said to be between two and three minutes — if you want to keep your audience to the end, try and keep your film within five minutes.

10. Keep viewer interest by making videos entertaining where possible, and using a variety of styles, e.g. expert interviews, site visits/tours, documentary, biographical, profiles etc.

11. Also be aware that different styles suit different audiences — just like writing an academic paper and rewording it for a more general audience. So, as mentioned below with regard to professional film-makers, try and do a shortened, more general version of your video, one for the academic (the specialist audience) and one for your stakeholders. It may be a good idea, especially if you are tight on budget or time, to make one film which can be used for both. This may require a little bit of extra planning and adjusting your own way of speaking, in particular using simple language to convey your message. Even academics will key into your video if you choose your words wisely and the content is relevant, has a story, is entertaining and visually pleasing or visceral (Box 12).

12. Attempt to make videos look as professional as possible, e.g. by adding introductory and end titles/credits.

13. Promote your video — just putting a video online won't necessarily get you any views. You need to integrate your video into your project's pathway to impact and think of ways to drive traffic to it. Simply embedding it in your project website won't help if you're not getting much traffic to your website. It can be particularly useful to invest in social media to drive traffic to online videos.

14. Have a go! Learn by doing it and get constructive feedback from your colleagues, but don't be too ambitious on your first attempt.

How do I get the most out of a professional film-maker?

Most of the points above apply when commissioning a video, except that the professionals will take care of much of this for you. Here are a few key pointers that can help you get the most out of working with film-makers:

- Although many film-makers will be able to help you refine your story, you will still need to provide them with the source material. Given that you understand your research best, you can often get much better results if you come to a film-maker with a few different ideas about how you might tell your story that they can then work with, rather than just sending them your latest paper or policy brief and hoping that they'll be able to come up with the story on their own.

- Think about who you'll need to interview and what locations you'd like to film — every extra day of filming on location adds to your budget, so if you can get everyone in the same place on your key location, you may be able to get filming days down to a minimum and save on costs.
- Once you know what you want from the film, you can negotiate a price — there will be an element of give and take, and you may have to scale back your ambitions depending on your budget.
- Ask the film-maker if they can provide a cut-down 'promotional' version of your film within the price or for a small additional fee — this can be an effective way of creating an additional version of your film that may be more relevant for a more generalist audience, to help expand who accesses your material.
- Make sure you check the draft version of your film carefully and provide detailed comments about things that need to be changed. Take time at this point and consult with the rest of your team, rather than going back and forth with lots of edits, or having colleagues objecting to content after the film is finished.

Publics/Stakeholder Analysis: worked example

This is based on a hypothetical stakeholder analysis developed for a project funded by the Swedish International Development and Co-operation Agency (Sida), led by the Regional Environment Centre in co-operation with local partner IUCN ROWA. The Water SUM project (http://www.watersum.rec.org/), for which this was developed, is using this template to train country teams how to conduct a stakeholder analysis in preparation for local water security action planning in collaboration with stakeholders. Column headings have been adapted for the purpose of this project. As long as you capture the level and nature of interest and influence in some way, you can add and remove columns from the matrix as necessary. However, remember that the more columns you add, the more time-consuming it will be to complete.

Name of organisation/group	Interest H/M/L	What are current levels of involvement in water management planning, and what aspects of local water security action planning (LWSAP) are they (likely to be) most interested in?	If involvement and/or interest is L/M, how might we motivate engagement with LWSAP? What benefits might they derive from being more involved in LWSAP?	Level of knowledge about water related issues H/M/L	Access to high-quality information about water-related issues H/M/L	Influence on water management H/M/L	Comments on influence (e.g. attitudes to water management planning, times or contexts in which they have more/less influence)	Any important relationships with other stakeholders? (e.g. conflicts/alliances)
Households	H	Involvement in water management planning varies significantly between households, but all households are water users, and are significantly affected by water management	N/A	L	L	L	None	Many households in the area rely on agriculture for at least part of their income, hence strong links with both types of farming stakeholder below
Farmers using	H	Farmers with land close to water sources growing crops that depend on	N/A	M	L	M	Those within the farming union and co-	Strong relationships with wider

Stakeholder		Interest				Engagement approach		
irrigated land		irrigation water are heavy water users and are significantly affected by water quality and quantity issues					operatives have a more organised, stronger voice	farming community including upland rain-fed farmers
Rain-fed upland farmers	L	Interested indirectly as householders or where they also own irrigated lowland fields, but otherwise not directly affected by changes in water flow or quality	L	L	L	Given low interest and influence, it is not a priority to engage with this group	None	Strong relationships with irrigated farming community, often through family ties
Farmer's Union	H	The Farmer's Union has been putting pressure on government for some time not to restrict access to irrigated water, and to invest in schemes to pipe water from other regions to this area	H	M	M	N/A	Despite having strong relationships with some politicians, the Farmer's Union has failed to achieve the objectives it has been campaigning for	Irrigated farmers are well represented in the Union, but upland farmers feel under-represented and membership from this group is much lower

Stakeholder		Description	Action				Impact	Notes
Local small businesses that depend on regular flows of clean water e.g. food and drink sector	H	Without adequate alternative supplies, problems with water quality and quantity are a major problem for some businesses in the area	N/A	L	L	L	None	Few small business have strong links with the government, farming or NGO communities, which reduces their influence
Multinational businesses	L	The local steel works is water-intensive, but is located upstream from most other water users, and so has preferential access to low flows, and has little interest in the problems this creates downstream, especially during drought years	Explore how more efficient water use might reduce costs and hence increase profits for the company. Look for evidence of failures to meet regulatory requirements to see if legal action could be taken. Explore potential for public campaign (including via print	L	H	H	Water use by this company is one of the key causes of low flows, and increased pollution levels	The CEO has married into a wealthy local family who have farming interests

			...and social media) to exert pressure on the company					
Government public health agencies	M	High interest in specific areas where pollution is leading to health problems and during drought years, but otherwise less directly interested in water management	N/A	L	M	L	There is a lack of communication between different government departments and agencies	Generally disconnected from other stakeholders affected by these issues
Environmental protection agencies from government	H	Statutory obligation to monitor and manage water resources	N/A	H	H	M	Due to limited resources, this agency has historically not been able to effect significant change in water resources management	There is conflict between these agencies and environmental NGOs who have been putting pressure on them to improve water managemen

IUCN water management project	H	High interest within the project team that is focusing on water management	N/A	H	H	M	At this point, the project is not well enough known to be able to estimate its influence, but if the project achieves its goals, then it will have been highly influential. Of course, if it does not achieve its goals, then its influence will have been low	Strong relationships with eNGOs and government (one of the only environmental organisations to have positive relationships with the environmental protection agency)
Other environmental NGOs	M	Other eNGOs are focusing on a wide range of topics, and do not have specific programmes relating to water management, however, they are indirectly interested when problems	N/A	M	M	L	Don't tend to work specifically on water management, so have relatively little influence over water	Involved in a number of long-standing conflicts over nature conservation and natural resource

		with water management compromise species and habitats that they are working on					management issues	management with the government
Local university	H	There is a strong research group focusing on Integrated Water Management who are collaborating with the IUCN project	N/A	H	H	L	The group has not engaged with stakeholders much in the past	Although links with other stakeholders are weak, the group is widely trusted by others

Example event facilitation plan

Aim: to identify organisations and groups with a stake in the future of sustainable food production in the UK which may be interested in co-producing knowledge with researchers from AFRP

09.45   Tea/coffee

10.00 Introduction and scoping

- Introductions
- Clarify the scope of the forum by geographical scale, sector etc. (where should our focus lie?)

10.20 Introduction to stakeholder analysis

- Introduction to stakeholder analysis and example using blank matrix on wall. Key points: defining interest and influence (both positive and negative), checking for redundant or missing columns
- Explain columns:
  - Name of organisation or group
  - Interest (H/M/L): how interested are they (likely to be) in the work?
  - Nature of interest: how do their interests intersect with the work, what are they likely to be most interested from the work?
  - Influence (H/M/L): how strongly might they be able to facilitate or block the work?
  - Comments on influence: why are they influential or not, and how could they help or block the project?
  - (to be completed at end): Steps to engagement — if low interest and high influence, how should we engage with them and incentivise their attendance?

10.30 Stakeholder analysis

- As a facilitated group discussion, list stakeholders organisations and groups in column A (Mark to facilitate, someone to scribe)

- Look through prompts to identify missing stakeholders, including marginalised groups and those who typically work on different spatial scales e.g. international organisations
- Facilitated group discussion to describe each stakeholder, column by column (participants adding additional comments via Post-it notes to ensure all knowledge is captured)

11.30 Break

11.45 Stakeholder analysis (continued)

- Individually complete the columns for all the remaining stakeholders, adding rows for additional stakeholders as they arise

13.00 Lunch

Plan A:

13:45 Checking the analysis

- All participants to check the work done by other participants, adding comments with Post-it notes where there is 'disagree' or 'don't understand'
- Facilitated discussion of key points that people feel should be discussed as a group about stakeholders where there is particular disagreement or confusion and resolve these where possible (accepting differing views where it is not possible to resolve differences)

14.30 Next steps

- Identify 4–8 trusted individuals who you can check the analysis with, trying to get as wide a spread of different interests as possible (to do this, it may be necessary to start with a longer list and then identify people who are likely to provide similar views to reduce the length of the list)
- Identify stakeholders with high influence but low current engagement and ask:
  - What could motivate greater engagement?
  - What initial steps could we take to engage with them?

15.30 Close

Plan B (if stakeholder analysis not complete before lunch):

13.30 Stakeholder analysis (continued)

- Complete the matrix

14.30 Checking the analysis

- All participants to check the work done by other participants, adding comments with Post-it notes where there is 'disagree' or 'don't understand'
- Facilitated discussion of key points that people feel should be discussed as a group about stakeholders where there is particular disagreement or confusion and resolve these where possible (accepting differing views where it is not possible to resolve differences)

15.00 Next steps

- Identify stakeholders with high influence but low current engagement and ask:
    - What could motivate greater engagement?
    - What initial steps could we take to engage with them?

16.00 Close

Materials

- Flip-chart paper
- Post-its
- OHP pens
- Marker pens
- Blu-Tack
- Pre-filled flip-chart paper
- Prompts to help identify stakeholders (photocopied to A3 if possible)
- A4 prompt sheet for facilitators showing common categories of stakeholder in case any have been omitted

Further reading

Here are a few of my favourite texts on research impact.

Books

Bastow, S., Dunleavy, P. and Tinkler, J. (2014) *The impact of the social sciences: How academics and their research make a difference.* Sage.

Boaz, A., Davies, H., Fraser, A. and Nutley, S. (2018) *What works now? Evidence-based policy and practice revisited.* Policy Press.

Denicolo, P. (Ed.) (2013) *Achieving impact in research.* Sage.

Derrick, G. (2018) *The evaluators' eye: Impact assessment and academic peer review.* Springer.

Dicks, L., Haddaway, N., Hernández-Morcillo, M., Mattsson, B., Randall, N., Failler, P., Ferretti, J., Livoreil, B., Saarikoski, H., Santamaria, L. and Rodela, R. (2017) *Knowledge synthesis for environmental decisions: an evaluation of existing methods, and guidance for their selection, use and development.* A report from the EKLIPSE project.

EU (2013) *Responsible Research and Innovation (RRI), Science and Technology: Summary.* Volume 401 of Special Eurobarometer, Publications Office, Brussels.

Useful articles

Berkes, F. (2009) Evolution of co-management: Role of knowledge generation, bridging organizations and social learning. *Journal of Environmental Management* 90: 1692–1702.

Boaz, A., Fitzpatrick, S., and Shaw, B. (2008) Assessing the impact of research on policy: A literature review. *Science and Public Policy* 36(4): 255–270.

Boaz, A., Locock, L., and Ward, V. (2015) Whose evidence is it anyway? *Evidence & Policy* 11: 145–148.

Cairney, P. and Oliver, K., 2017. Evidence-based policymaking is not like evidence-based medicine, so how far should you go to bridge the divide between evidence and policy? *Health Research Policy and Systems* 15:35.

Contandriopoulos, D., Lemire, M., Denis, J.-L., and Tremblay, E. (2010) Knowledge exchange processes in organizations and policy arenas: A narrative systematic review of the literature. *Milbank Quarterly* 88: 444–483.

de Vries, J.R., Roodbol-Mekkes, P., Beunen, R., Lokhorst, A.M. and Aarts, N., 2014. Faking and forcing trust: The performance of trust and distrust in public policy. *Land Use Policy* 38: 282-289.

de Vries, H., Tummers, L. and Bekkers, V. (2018) The diffusion and adoption of public sector innovations: A meta-synthesis of the literature. *Perspectives on Public Management and Governance*.

Greenhalgh, T. and Fahy, N. (2015) Research impact in the community-based health sciences: An analysis of 162 case studies from the 2014 UK Research Excellence Framework. *BMC Medicine* 13: 232–244.

Heaney, M.T. (2006) Brokering health policy: Coalitions, parties, and interest group influence. *Journal of Health Politics, Policy & Law* 31: 887–944.

Lavis, J.N., Oxman, A.D., Moynihan, R. and Paulsen, E.J. (2008) Evidence-informed health policy 1— Synthesis of findings from a multi-method study of organizations that support the use of research evidence. *Implementation Science* 3: 7.

Marshall, N., Adger, N., Attwood, S., Brown, K., Crissman, C., Cvitanovic, C., et al. (2017) Empirically derived guidance for social scientists to influence environmental policy. *PLoS ONE* 12(3): e0171950.

Meagher, L.R. and Martin, U. (2017) Slightly dirty maths: The richly textured mechanisms of impact. *Research Evaluation* 26: 15-27.

Oancea, A. (2013) Interpretations of research impact in seven disciplines. *European Educational Research Journal* 12: 242–250.

Oliver, K., Innvaer, S., Lorenc, T., Woodman, J. and Thomas, J. (2014) A systematic review of barriers to and facilitators of the use of evidence by policymakers. *BMC Health Services Research* 14: 2.

Owen, C., Hemmings, L. and Brown, T. (2009) Lost in translation: Maximizing handover effectiveness between paramedics and receiving staff in the emergency department. *Emergency Medicine Australasia* 21, 102e107.

Owen, R., Macnaghten, P. and Stilgoe, J. (2012) Responsible research and innovation: From science in society to science for society, with society. *Science and Public Policy* 39: 751–760.

Pardoe, S. (2014) Research impact unpacked? A social science agenda for critically analyzing the discourse of impact and informing practice. *SAGE Open* 4: 2.

Phillipson, J., Lowe, P., Proctor, A. and Ruto, E. (2012) Stakeholder engagement and knowledge exchange in environmental research. *Journal of Environmental Management* 95: 56–65.

Rivera, S.C., Kyte, D.G., Aiyegbusi, O.L., Keeley, T.J. and Calvert, M.J. (2017) Assessing the impact of healthcare research: A systematic review of methodological frameworks. *PLoS Medicine* 14 (8), p.e1002370.

Smith, K.E. and Stewart, E. (2017) We need to talk about impact: Why social policy academics need to engage with the UK's research impact agenda. *Journal of Social Policy* 46, 1: 109–127.

Ward, V., Smith, S.O., House, A. and Hamer, S. (2012) Exploring knowledge exchange: A useful framework for practice and policy. *Social Science and Medicine* 74: 297–304.

By the author

Here are some of my attempts to make contributions in this field so far. Although they have mainly been written in an environmental context, the insights from many of these papers are generalisable. You can find links to each paper on my website (www.profmarkreed.com). Contact me if you don't have access to any of the journals and I'll send you a copy.

Reed, M.S., Ferre, M., Martin-Ortega, J. Blanche, R., Dallimer, M. and Holden, J. (2018) Evaluating and evidencing research impact: A literature review. *Plos One.*

Wyborn, C., Reed, M.S. et al. (2018) Understanding the impacts of synthesis research. *Environmental Science and Policy.*

Reed, M.S., Bryce, R., and Machen, R. (2018) Pathways to policy impact: A new approach for planning and evidencing research impact. *Evidence & Policy.*

Reed, M.S., Duncan, S., Manners, P., Pound, D., Armitage, L., Frewer, L., Thorley, C. and Frost, B. (2018) A common standard for the evaluation of public engagement with research. *Research For All.*

Chubb, J., and Reed, M.S. (2018). The politics of research impact: Implications for research funding, motivation and quality. *British Politics.*

Reed, M.S., Vella, S., Challies, E., de Vente, J., Frewer, L., Hohenwallner-Ries, D., Huber, T., Neumann, R.K., Oughton, E.A., Sidoli del Ceno, J. and van Delden, H. (2017) A theory of participation: What makes stakeholder and public engagement in environmental management work? *Restoration Ecology.*

Reed MS (2017) *The Productive Researcher.* Fast Track Impact.

de Vente, J., Reed, M.S., Stringer, L.C., Valente, S. and Newig, J. (2016) How does the context and design of participatory decision-making processes affect their outcomes? Evidence from sustainable land management in global drylands. *Ecology & Society* 21(2): 24.

Reed, M.S. and Curzon, R. (2015) Stakeholder mapping for the governance of biosecurity: A literature review. *Journal of Integrative Environmental Sciences* 12: 15–38.

Reed, M.S., Stringer, L.C., Fazey, I., Evely, A.C. and Kruijsen, J. (2014). Five principles for the practice of knowledge exchange in environmental management. *Journal of Environmental Management* 146: 337–345.

Fazey, I., Bunse, L., Msika, J., Pinke, M., Preedy, K., Evely, A.C., Lambert, E., Hastings, E., Morris, S., and Reed, M.S. (2014) Evaluating knowledge exchange in interdisciplinary and multi-stakeholder research. *Global Environmental Change* 25: 204–220.

Reed, M.S., Fazey, I., Stringer, L.C., Raymond, C.M., Akhtar-Schuster, M., Begni, G., Bigas, H., Brehm, S., Briggs, J., Bryce, R., Buckmaster, S.,

Chanda, R., Davies, J., Diez, E., Essahli, W., Evely, A., Geeson, N., Hartmann, I., Holden, J., Hubacek, K., Ioris, I., Kruger, B., Laureano, P., Phillipson, J., Prell, C., Quinn, C.H., Reeves, A.D., Seely, M., Thomas, R., van der Werff Ten Bosch, M.J., Vergunst, P. and Wagner, L. (2013) Knowledge management for land degradation monitoring and assessment: An analysis of contemporary thinking. *Land Degradation & Development* 24: 307–322.

Reed, M.S., Bonn, A., Broad, K., Burgess, P., Fazey, I., Fraser, E.D.G., Hubacek, K., Nainggolan, D., Roberts, P., Quinn, C.H., Stringer, L.C., Thorpe, S., Walton, D.D., Ravera, F. and Redpath, S. (2013) Participatory scenario development for environmental management: A methodological framework. *Journal of Environmental Management* 128: 345–362.

Fazey, I., Evely, A.C., Reed, M.S., Stringer, L.C., Kruijsen, J., White, P.C.L., Newsham, A., Jin, L., Cortazzi, M., Phillopson, J., Blackstock, K., Entwhistle, N., Sheate, W., Armstrong, F., Blackmore, C., Fazey, J., Ingram, J., Gregson, J., Lowe, P., Morton, S. and Trevitt, C. (2012) Knowledge exchange: A research agenda for environmental management. *Environmental Conservation* 40: 19–36.

Prell, P., Reed, M.S., Racin, L. and Hubacek, K. (2010) Competing structures, competing views: The role of formal and informal social structures in shaping stakeholder perceptions. *Ecology & Society* 15(4): 34.

Reed, M.S., Evely, A.C., Cundill, G., Fazey, I., Glass, J., Laing, A., Newig, J., Parrish, B., Prell, C., Raymond, C. and Stringer, L.C. (2010) What is social learning? *Ecology & Society* 15 (4): r1.

Raymond, C.M., Fazey, I., Reed, M.S., Stringer, L.C., Robinson, G.M. and Evely, A.C. (2010) Integrating local and scientific knowledge for environmental management: From products to processes. *Journal of Environmental Management* 91: 1766–1777.

Prell, C., Hubacek, K., and Reed, M.S. (2009) Social network analysis and stakeholder analysis for natural resource management. *Society & Natural Resources* 22: 501–518.

Reed, M.S., Graves, A., Dandy, N., Posthumus, H., Hubacek, K., Morris, J., Prell, C., Quinn, C.H. and Stringer, L.C. (2009) Who's in and why? Stakeholder analysis as a prerequisite for sustainable natural resource management. *Journal of Environmental Management* 90: 1933–1949.

Prell, C., Hubacek, K., Quinn, C. and Reed M.S. (2008) 'Who's in the network?' When stakeholders influence data analysis. *Systemic Practice and Action Research* 21: 443–458.

Reed, M.S. (2008) Stakeholder participation for environmental management: A literature review. *Biological Conservation* 141: 2417–2431.

Prell, C., Hubacek, K., Reed, M.S., Burt, T.P., Holden, J., Jin, N., Quinn, C., Sendzimir, J. and Termansen, M. (2007) If you have a hammer everything looks like a nail: 'Traditional' versus participatory model building. *Interdisciplinary Science Reviews* 32: 1–20.

Reed, M.S., Dougill, A.J. and Taylor, M.J. (2007) Integrating local and scientific knowledge for adaptation to land degradation: Kalahari rangeland management options. *Land Degradation & Development* 18: 249–268.

Fraser, E.D.G., Dougill, A.J., Mabee, W., Reed, M.S. and McAlpine, P. (2006) Bottom up and top down: Analysis of participatory processes for sustainability indicator identification as a pathway to community empowerment and sustainable environmental management. *Journal of Environmental Management* 78: 114–127.

Dougill, A.J., Fraser, E.D.G., Holden, J., Hubacek, K., Prell, C., Reed, M.S., Stagl, S.T. and Stringer, L.C. (2006) Learning from doing participatory rural research: Lessons from the Peak District National Park. *Journal of Agricultural Economics* 57: 259–275.

Stringer, L.C., Prell, C., Reed, M.S., Hubacek, K., Fraser, E.D.G. and Dougill, A.J. (2006) Unpacking 'participation' in the adaptive management of socio-ecological systems: A critical review. *Ecology & Society* 11: 39

Acknowledgements

First edition acknowledgements

This book is based on research funded by the UK's Economic and Social Research Council (ESRC), Biotechnology and Biological Sciences Research Council (BBSRC) and the Natural Environment Research Council (NERC), with additional funding provided by the Scottish Government and the Department for Environment, Food and Rural Affairs (under the Rural Economy and Land Use programme and the Living With Environmental Change programme, LWEC) (the Sustainable Learning project), and the British Academy (the Involved project). I am indebted to the teams of researchers I worked with in both of these projects, in particular Ana Attlee, Ioan Fazey and Lindsay Stringer in the Sustainable Learning project, and Joris de Vente, Lindsay Stringer, Jens Newig and Sandra Valente in the Involved project. Thanks to Joanneke Kruisjen for valuable inputs to the thinking behind the key paper that emerged from the Sustainable Learning project. This book started life as LWEC's Knowledge Exchange Guidelines, and was then developed into a training manual I co-authored with Ana Attlee, to accompany the training course we developed together. Thanks to Susan Ballard (LWEC), Faith Culshaw (NERC), John Holmes (NERC/LWEC Fellow) and Barry Hague (editor) for their inputs to the LWEC KE Guidelines from which the principles in this book were derived.

Thanks to the knowledge exchange and impact experts who helped co-design the research that led to this book: Joanneke Kruijsen (School of Engineering, Robert Gordon University), Piran White (Environment Department, University of York), Andrew Newsham (School of African and Oriental Studies (SOAS), University of London), Lixian Jin (School of Allied Health Sciences, de Montford University), Martin Cortazzi (Centre for Applied Linguistics, University of Warwick), Jeremy Phillipson (Department of Agriculture, Food and Rural Development, Newcastle University), Kirsty Blackstock (Social, Economic and Geographical Sciences, James Hutton Institute), Noel Entwistle (School of Education, University of Edinburgh), William Sheate (Centre for Environmental Policy, Imperial College London), Fiona Armstrong (Head of Knowledge Exchange, Economic and Social Research Council), Chris Blackmore (Open Systems Research Group, Faculty of

Mathematics, Open University), John Fazey (Educationalist, Ty'n Y Caeau Consultants), Julie Ingram (Countryside and Community Research Institute, University of Gloucestershire), Jon Gregson (Institute of Development Studies), Philip Lowe (UK Research Councils' Rural Economy and Land Use programme and Newcastle University), Sarah Morton (Centre for Research on Families and Relationships, University of Edinburgh) and Chris Trevitt (College of Law, Australian National University).

I am grateful to the following people who kindly gave their time to review and give feedback on the first edition of this book: Jenny Ames (Associate Dean (Research and Innovation) and Professor of Food and Nutritional Sciences in the Faculty of Health and Applied Sciences, University of the West of England), Gabriele Bammer (Professor, Research School of Population Health, College of Medicine, Biology and Environment, Australian National University), Mahantesh Biradar (Institute of Biomedical Sciences at Academia Sinica, Taipei, National Yang Ming University, Taiwan), Rebecca Colvin (PhD student, University of Queensland, Australia), Grant Campbell (PhD student, Cranfield University and James Hutton Institute), Georgina Cosma (Senior Lecturer in Computer Science, School of Science and Technology, Nottingham Trent University), Ged Hall (Innovation and Enterprise Senior Training and Development Officer, University of Leeds), Dr Ingrid Hanson (English literature scholar and author of *William Morris and the Uses of Violence, 1856-1890*), Steven Hill (Head of Research Policy, Higher Education Funding Council for England, HEFCE), Anne Liddon (Science Communications Manager, Newcastle University), Paul Manners (Director, UK National Co-ordinating Centre for Public Engagement), Catherine Manthorpe (Head of the Research Office, University of Hertfordshire), Rosmarie Katrine Neumann (PhD student, Newcastle University), Katharine Reibig (Researcher Development Policy Officer, University of Stirling), Elizabeth Stokoe (Professor of Social Interaction, Associate Dean for Research, School of Social, Political and Geographical Sciences, Loughborough University), and Steven Vella (PhD student, Birmingham City University and Newcastle University).

I am grateful to Ana Attlee for inspiring me to start on the journey that eventually led to this book while working as a post-doctoral researcher on the Sustainable Learning project, and for the subsequent contributions and feedback on drafts of our training manual, from which this book was developed. Thanks also to Ana for teaching me most of what I know about social media. Thanks to Clare Gately (my writing buddy) for valuable feedback and encouragement. Thanks to Roz Taylor from Elevator UK for advising me on marketing and business strategy, and for empowering me to turn my fears into enthusiasm. Thanks to Rosmarie Katrin Neumann for helping draft Chapter 21 and Sarah Buckmaster for research on HEFCE's impact case study database in Chapter 20. Thanks to Simon Maxell, Colin Smith, Catherine Duggan and Maggie Charnley from the UK Government's Department for Environment, Food & Rural Affairs (Defra) and Rosmarie Katrin Neumann for constructive feedback and edits on Chapter 20. Thanks to Emma Leech (Director of Marketing and Advancement at Loughborough University) for telling me about the problems associated with Advertising Value Equivalency and pointing me to CIPR resources. Some of the suggestions I've made in Chapters 15 and 16 are based on what I learned from training by Diana Pound from Dialogue Matters. Box 10 is based on questions designed by Ben Fuchs and Maggie Buxton from Cohesive ID. Thanks to Steven Vella for finding time to give feedback on this book during the hard grind of writing up his PhD, and for inputs to various chapters (in particular advice on film-making). Thanks to Sarah Lyth for showing me where to get amazing copyright-free photography and to Joyce Reed for additional photography.

Thanks to everyone at Birmingham City University, in particular Hanifa Shah, for the creative freedom and inspirational leadership that enabled me to chase my dreams and write the first edition of this book. I wouldn't have been able to write any of this if it were not for Philip Lowe, Jeremy Philipson, Anne Liddon and the rest of the RELU family who gave me my first big break, some of the most terrifying early tastes of working with policy-makers, and the support and confidence I needed to pursue impact. Winning the RELU impact prize is still the proudest moment of my career. Thanks to Marianne Noble for proof-reading the book. Finally, thanks to Madie Whittaker for organising my life so I could write this book, and thanks to Joyce, my wife, for cutting through my wordiness and inspiring me every day to do what I can to make the world a better place.

Additional acknowledgements for the 2nd Edition

Chapter 22 was the hardest to write of all the new material, and I am grateful to various colleagues who helped me with it. Thanks to Marie Ferre and Julia Martin-Ortega for input to this chapter as we wrote a review paper on the same topic in parallel with its development. Thanks to Sally-Anne Whiteman for inspiring and helping to write the section on 'evidencing international policy impact' in this chapter. Thanks to Sophie Duncan and Paul Manners from the National Co-ordinating Centre for Public Engagement, Diana Pound and Lucy Armitage from Dialogue Matters, Lynn Frewer from Newcastle University and Charlotte Thorley and Bryony Frost from Queen Mary University of London, who co-authored 'A common standard for the evaluation of public engagement with research' with me in *Research For All*, which I drew upon for part of the 'what should I evaluate?' section. That work was funded by the Higher Education Funding Council for England's Higher Education Innovation Fund and the Wellcome Trust's Institutional Strategic Support Fund, awarded to Queen Mary University of London, and was commissioned by Peter McOwan, Vice Principal (Public Engagement and Student Enterprise) at QMUL. Thanks to Marianne Noble for painstakingly proofing the second edition, and offering many useful suggestions that clarified the meaning of the text. Finally, the reason the second edition looks so much nicer than the first is that the whole thing was redesigned with loving care by Anna Sutherland from Fast Track Impact.